Random Motions in Markov and Semi-Markov Random Environments 2

Series Editor
Nikolaos Limnios

Random Motions in Markov and Semi-Markov Random Environments 2

High-dimensional Random Motions and Financial Applications

Anatoliy Pogorui
Anatoliy Swishchuk
Ramón M. Rodríguez-Dagnino

iSTE

WILEY

First published 2021 in Great Britain and the United States by ISTE Ltd and John Wiley & Sons, Inc.

ISTE Ltd
27-37 St George's Road
London SW19 4EU
UK

www.iste.co.uk

John Wiley & Sons, Inc.
111 River Street
Hoboken, NJ 07030
USA

www.wiley.com

Library of Congress Control Number: 2020946634

British Library Cataloguing-in-Publication Data
A CIP record for this book is available from the British Library
ISBN 978-1-78630-706-4

MIX
Paper from responsible sources
FSC® C013604

Contents

Preface ix

Acknowledgments xiii

Introduction xv

Part 1. Higher-dimensional Random Motions and Interactive Particles 1

Chapter 1. Random Motions in Higher Dimensions 3

1.1. Random motion at finite speed with semi-Markov switching directions process 5
1.1.1. Erlang-K-distributed direction alternations 7
1.1.2. Some properties of the random walk in a semi-Markov environment and its characteristic function 15
1.2. Random motion with uniformly distributed directions and random velocity 17
1.2.1. Renewal equation for the characteristic function of isotropic motion with random velocity in a semi-Markov media 17
1.2.2. One-dimensional case 20
1.2.3. Two-dimensional case 23
1.2.4. Three-dimensional case 23
1.2.5. Four-dimensional case 31
1.3. The distribution of random motion at non-constant velocity in semi-Markov media 32
1.3.1. Renewal equation for the characteristic function 34
1.3.2. Two-dimensional case 35
1.3.3. Three-dimensional case 37

1.3.4. Four-dimensional case 40
1.4. Goldstein–Kac telegraph equations and random flights in higher dimensions 43
1.4.1. Preliminaries about our modeling approach 45
1.4.2. Two-dimensional case 48
1.4.3. Three-dimensional case 51
1.4.4. Five-dimensional case 59
1.5. The jump telegraph process in $\mathbb{R}^n$ 62
1.5.1. The jump telegraph process in $\mathbb{R}^3$ 63
1.5.2. Conclusions and final remarks 64

Chapter 2. System of Interactive Particles with Markov and Semi-Markov Switching 67

2.1. Description of the Markov model 68
2.1.1. Distribution of the first meeting time of two telegraph processes 69
2.1.2. Estimate of the number of particle collisions 74
2.1.3. Free path times of a family of particles 76
2.1.4. Estimation of the number of particle collisions for systems with boundaries 78
2.1.5. Estimation of the number of particle collisions for systems without boundaries 83
2.2. Interaction of particles governed by generalized integrated telegraph processes: a semi-Markov case 87
2.2.1. Laplace transform of the distribution of the first collision of two particles 88
2.2.2. Semi-Markov case 91
2.2.3. Distribution of the first collision of two particles with finite expectation 95

Part 2. Financial Applications 99

Chapter 3. Asymptotic Estimation for Application of the Telegraph Process as an Alternative to the Diffusion Process in the Black–Scholes Formula 101

3.1. Asymptotic expansion for the singularly perturbed random evolution in Markov media in the case of disbalance 101
3.2. Application: Black–Scholes formula 106

Chapter 4. Variance, Volatility, Covariance and Correlation Swaps for Financial Markets with Markov-modulated Volatilities 111

4.1. Volatility derivatives 111
4.1.1. Types of volatilities 111
4.1.2. Models for volatilities 113

4.1.3. Variance and volatility swaps 115
4.1.4. Covariance and correlation swaps 116
4.1.5. A brief literature review 117
4.2. Martingale representation of a Markov process 118
4.3. Variance and volatility swaps for financial markets with Markov-modulated stochastic volatilities 122
4.3.1. Pricing variance swaps 124
4.3.2. Pricing volatility swaps 124
4.4. Covariance and correlation swaps for two risky assets for financial markets with Markov-modulated stochastic volatilities 128
4.4.1. Pricing covariance swaps 128
4.4.2. Pricing correlation swaps 130
4.4.3. Correlation swap made simple 130
4.5. Example: variance, volatility, covariance and correlation swaps for stochastic volatility driven by two state continuous Markov chain 132
4.6. Numerical example 134
4.6.1. S&P 500: variance and volatility swaps 134
4.6.2. S&P 500 and NASDAQ-100: covariance and correlation swaps . . 135
4.7. Appendix 1 138
4.7.1. Correlation swaps: first-order correction 138

Chapter 5. Modeling and Pricing of Variance, Volatility, Covariance and Correlation Swaps for Financial Markets with Semi-Markov Volatilities 143

5.1. Introduction 143
5.2. Martingale representation of semi-Markov processes 148
5.3. Variance and volatility swaps for financial markets with semi-Markov stochastic volatilities 151
5.3.1. Pricing of variance swaps 153
5.3.2. Pricing of volatility swaps 155
5.3.3. Numerical evaluation of variance and volatility swaps with semi-Markov volatility 158
5.4. Covariance and correlation swaps for two risky assets in financial markets with semi-Markov stochastic volatilities 159
5.4.1. Pricing of covariance swaps 160
5.4.2. Pricing of correlation swaps 162
5.5. Numerical evaluation of covariance and correlation swaps with semi-Markov stochastic volatility 164

5.6. Appendices 165
5.6.1. Appendix 1. Realized correlation: first-order correction 165
5.6.2. Appendix 2. Discussions of some extensions 169

References 177

Index 191

Summary of Volume 1 193

Preface

Motion is an essential element in our daily life. Thoughts related to motion can be found in the ancient Greek philosophers; however, we have to look several centuries ahead for relevant mathematical models. Galileo Galilei, Isaac Newton and Johannes Kepler, between the years 1550–1650, made remarkable advances in the construction of mathematical models for deterministic motion. Further advances in this line were made by Leonhard Euler and William Rowan Hamilton, and, in 1778, Joseph-Louis Lagrange proposed a new formulation of classical mechanics. This new formulation is based on the optimization of energy functionals, and it allows us to solve more sophisticated motion problems in a more systematic manner. This new approach to mechanics is the basis for dealing with modern quantum mechanics and the physics of high-energy particles. Albert Einstein, in 1905, through his special theory of relativity, introduced fine corrections for high velocities, close to the maximum speed of light. All of these fundamental mathematical models deal with many sophisticated motions such as that of astronomical objects, satellites (natural and artificial), particles in intense electromagnetic fields, particles under gravitational forces, a better understanding of light, and so on. However, even though these might be very complicated problems, we should mention that all of them have deterministic paths of motion.

In 1827, the Scottish botanist Robert Brown described a special kind of *random* motion produced by the interaction of many particles, while looking at pollen of the plant *Clarkia pulchella* immersed in water, using a microscope, and it was recognized that this type of motion could not be fully explained by modeling the motion of each particle or molecule. Albert Einstein, 78 years later, in 1905, published a seminal paper where he modeled the motion of the pollen as being moved by individual water molecules. In this work, the diffusion equation was introduced as a convenient mathematical model for this random phenomenon. A related model in this direction was presented in 1906 by Marian Smoluchowski, and the experimental verification was done by Jean Baptiste Perrin in 1908.

A similar model for Brownian motion was proposed in 1900 by Louis Bachelier in his PhD thesis entitled *The Theory of Speculation*, where he presented a stochastic analysis for valuing stock options in financial markets. This novel application of a stochastic model faced criticism at the beginning, but Bachelier's instructor Henri Poincaré was in full support of this visionary idea. This fact shows a close relationship between models to explain random phenomena in physics (statistical mechanics) and in financial analysis, and also in many other areas.

A notable contribution of the American mathematician Norbert Wiener was to establish the mathematical foundations for Brownian motion, and for that reason it is also known as the Wiener process. Great mathematicians such as Paul Lévy, Andrey Kolmogorov and Kiyosi Itô, among many other brilliant experts in the new field of probability and stochastic processes, set the basis of these stochastic processes. For example, the famous Black–Scholes formula in financial markets is based on both diffusion processes and Itô's ideas.

In spite of its success in modeling many types of random motion and other random quantities, the Wiener process has some drawbacks when capturing the physics of many applications. For instance, the modulus of velocity is almost always infinite at any instant in time, it has a free path length of zero, the path function of a particle is almost surely non-differentiable at any given point and its Hausdorff dimension is equal to 1.5, i.e. the path function is fractal. However, the actual movement of a physical particle and the actual evolution of share prices are barely justified as fractal quantities. Taking into account these considerations, in this book we propose and develop other stochastic processes that are close to the actual physical behavior of random motion in many other situations. Instead of the diffusion process (Brownian motion), we consider telegraph processes, Poisson and Markov processes and renewal and semi-Markov processes.

Markov (and semi-Markov) processes are named after the Russian mathematician Andrey Markov, who introduced them in around 1906. These processes have the important property of changing states under certain rules, i.e. they allow for abrupt changes (or switching) in the random phenomenon. As a result, these models are more appropriate for capturing random jumps, alternate velocities after traveling a certain random distance, random environments through the formulation of random evolutions, random motion with random changes of direction, interaction of particles with non-zero free paths, reliability of storage systems, and so on.

In addition, we will model financial markets with Markov and semi-Markov volatilities as well as price covariance and correlation swaps. Numerical evaluations of variance, volatility, covariance and correlations swaps with semi-Markov volatility are also presented. The novelty of these results lies in the pricing of volatility swaps in the closed form, and the pricing of covariance and correlation swaps in a market with two risky assets.

Anatoliy POGORUI
Zhytomyr State University, Ukraine

Anatoliy SWISHCHUK
University of Calgary, Canada

Ramón M. RODRÍGUEZ-DAGNINO
Tecnologico de Monterrey, Mexico

October 2020

Acknowledgments

Anatoliy Pogorui was partially supported by the State Fund for Basic Research (Ministry of Education and Science of Ukraine 20.02.2017 letter no. 12).

Anatoliy Pogorui

I would like to thank NSERC for its continuing support, my research collaborators, and my current and former graduate students. I also give thanks to my family for their inspiration and unconditional support.

Anatoliy Swishchuk

I would like to thank Tecnologico de Monterrey for providing me the time and support for these research activities. I also appreciate the support given by Conacyt through the project no. SEP-CB-2015-01-256237. The time and lovely support given to me by my wife Saida, my three daughters Dunia, Melissa and R. Melina, as well as my son Ramón Martín, are invaluable.

Ramón M. Rodríguez-Dagnino

Introduction

I.1. Overview

The theory of dynamical systems is one of the fields of modern mathematics that is under intensive study. In the theory of stochastic processes, researchers are actively studying dynamical systems, operating under the influence of random factors. A good representative of such systems is the theory of random evolutions. The first results within this field were obtained by Goldstein (1951) and Kac (1974), who studied the movement of a particle on a line with a speed that changes its sign under the Poisson process. Subsequently, this process was called the telegraph process or the Goldstein–Kac process. Further developments of this theory have been presented in the works of Griego and Hersh (1969, 1971), Hersh and Pinsky (1972), and Hersh (1974, 2003), which gave a definition of stochastic evolutions in a general setting.

Important advances in the theory of stochastic evolutions have been made in the formulation of limit theorems and their refinement; these consist of obtaining asymptotic expansions. These problems are studied in the works of Korolyuk and Turbin (1993), Skorokhod (1989), Turbin (1972, 1981), Korolyuk and Swishchuk (1986), Korolyuk and Limnios (2009, 2005), Shurenkov (1989, 1986), Dorogovtsev (2007b), Anisimov (1977), Girko (1982), Hersh and Pinsky (1972), Papanicolaou (1971a,b), Kertz (1978), Watkins (1984, 1985), Balakrishnan *et al.* (1988), Yeleyko and Zhernovyi (2002), Pogorui (1989, 1994, 2009a) and Pogorui and Rodríguez-Dagnino (2006, 2010a).

Among the many methods in finding limiting theorems, we should mention a class that could be called an asymptotic average scheme. By using these methods, a semi-Markov evolution can be reduced to a random evolution in a lumping state space with Markov switching, and is thus studied as a Markov scheme. Most of the results in this field are presented in the following books and papers: Korolyuk and Swishchuk (1995), Korolyuk and Korolyuk (1999), Korolyuk and Turbin (1993),

Korolyuk and Limnios (2005), Turbin (1972), Pogorui (2004, 2012a), and Rodríguez-Said *et al.* (2007). Furthermore, this theory was also developed in the study of asymptotic expansions for functionals of random evolution in the phase averaging and diffusion approximation. This topic was the main subject of the following works: Korolyuk and Limnios (2009, 2004), Turbin (1981), Samoilenko (2005), Albeverio *et al.* (2009), Pogorui (2010a), Pogorui and Rodríguez-Dagnino (2010a), and Nischenko (2001).

The asymptotic average scheme has been applied to a semi-Markov evolution for the computation of the effectiveness of a multiphase system with a couple of storage units by Pogorui (2003, 2004), Pogorui and Turbin (2002), and Rodríguez-Said *et al.* (2007).

Research on random evolutions has also been carried out by applying martingale methods. It seems that the founders of this approach were Stroock and Varadhan (1969, 1979), but further developments can be found in the works of Skorokhod (1989), Pinsky (1991), Korolyuk and Korolyuk (1999), Sviridenko (1986), Swishchuk (1989), Hersh and Papanicolaou (1972), Iksanov and Rösler (2006), and Griego and Korzeniowski (1989).

In addition to the successful development of abstract stochastic evolutions in the decade 1980–1990, several scholars studied various generalizations of the Goldstein–Kac telegraph process to multidimensional spaces. In connection to this, the results of Gorostiza (1973), Gorostiza and Griego (1979), Orsingher (1985), Orsingher and Somella (2004), Turbin (1998), Orsingher and Ratanov (2002), Samoilenko (2001) and Lachal (2006) should be noted. In most of these works, the authors considered a finite number of directions of a particle movement and obtained differential equations for the probability density functions. In the works of Pogorui (2007) and Pogorui *et al.* (2014), the authors proposed a method for solving such equations by using monogenic functions, after associating a particular commutative algebra with them.

Masoliver *et al.* (1993a) studied the telegraph process with reflecting or partially reflecting boundaries and its distribution in a fixed interval of time. In papers by Pogorui (2005, 2006), and Pogorui and Rodríguez-Dagnino (2006, 2010b), the authors also study the stationary distribution of some Markov and semi-Markov evolution with delaying boundaries.

De Gregorio *et al.* (2005), and Stadje and Zacks (2004) considered the generalizationof the telegraph process on a line, where there is a discrete set of particle velocities, then at Poisson epochs a velocity is chosen from this set.

Pogorui (2010b), Pogorui and Rodríguez-Dagnino (2009b) and Samoilenko (2002) investigated fading evolutions, where the velocity of a particle tends to zero as the number of switches grows at infinite.

Pogorui and Rodríguez-Dagnino (2005) also studied a generalization of the telegraph process to the case of Erlang interarrivals between successive switches of particle velocities. For such processes, a differential equation for the probability density function (pdf) of the particle position on the line was obtained. In addition, in Pogorui (2011) a method for solving such equations, by using monogenic functions associated with this equation on a commutative algebra, was developed.

Orsingher and De Gregorio (2007), Stadje (2007) and others studied the motion of a particle in multidimensional spaces with constant absolute velocity and directions uniformly distributed on a unit sphere that change at Poisson points. The authors obtained explicit formulas for the position of the particle distributions for two- and four-dimensional space and investigated the "explosive effect" for the pdf of the position of a particle that approaches the singularity sphere for the plane and the three-dimensional space.

In recent years, much attention has been paid to the Pearson random walk with Gamma distributed steps and associated walks of Pearson–Dirichlet type, whose steps have the Dirichlet distribution. Franceschetti (2007) developed explicit formulas for the conditional pdf of the position of a particle at any number of steps for a walk in $\mathbb{R}^n$ $(n = 1, 2)$ with uniformly distributed directions and steps with the Dirichlet distribution with parameter $q = 1$. Beghin and Orsingher (2010a) obtained an expression for the conditional distribution of the position of a particle of the Pearson–Dirichlet walk with parameter $q = 2$ in the plane. Le Caër (2010, 2011) generalized these results to the case of the multidimensional Pearson–Dirichlet random walk with arbitrary parameter q, where the author introduces the concept of a "Hyperspherical Uniform" (HU) random walk. The HU walk is a motion, the endpoint distribution of which is identical to the distribution of the projection in the walk space of a point, with a position vector, randomly chosen on the surface of the unit hypersphere of some hyperspace in higher dimensions. By using properties of the HU random walk, Le Caër found walks for which the conditional probability density can be expressed in a closed form. We should also mention the recent papers by De Gregorio (2014) and Letac and Piccioni (2014), where they obtained a generalization and simplification of proofs of the results stated by Le Caër.

In all of the above-mentioned papers, the authors study the conditional distributions of the particle position at renewal epochs of the switching direction process. For a non-Markov switching process, it is possible to find the corresponding results. The mathematical technique is to replace the distribution of the non-Markov process by the Markov chain walk embedded in this process. By using this approach, Pogorui and Rodríguez-Dagnino (2011) obtained a recursive expression for the conditional characteristic functions of a random walk with Erlang switching, considering a non-Markov switching process. Namely, they studied the changes in the conditional characteristic functions of the particle position, not only at instants of the direction changing, but at all Poisson times. Pogorui (2011) studied the particular case of the Erlang distributed stay of the switching process in the states. Further results in this direction were obtained by Pogorui and Rodríguez-Dagnino (2012) where the authors also studied multidimensional random motion at random velocities. For some distributions of random velocity, they observed an "explosive effect" for the pdf of the position of a particle when it is approaching the singularity sphere in four-dimensional space.

Other directions of random walk theory, which have been studied intensively during recent years, are the fractal Brownian motion and the fractal generalization of the telegraph process. These processes have been studied by Qian *et al.* (1998), Cahoy (2007), Beghin and Orsingher (2010b), Orsingher and Beghin (2009), D'Ovidio *et al.* (2014), and others.

The set of particles with interaction, where each particle moves on a line according to a telegraph process, up to collision with another particle, was studied by Pogorui (2012b). During the collision, the particles exchange momentums. In this book, the author calculates the distribution of time of the first collision for two telegraph particles that started simultaneously from different points on a line and investigates the limit of this distribution under Kac's condition. The author also investigates the system of particles with Markov switching, which is bounded with reflecting boundaries. The distribution for the position of particles of the system in a fixed time was also obtained. The limiting properties of these distributions and an estimate of the number of collisions in the system with reflecting boundaries, as well as without them, are also studied. Such a system of particles can be interpreted as a model of one-dimensional gas and it is a kind of one-dimensional generalization of the deterministic models of gas, of the billiard type, that were studied by Kornfeld *et al.* (1982), for example. The velocity of particles in these models is considered to be finite. This is a major difference from systems where the position of a particle is described by a diffusion process, such as in Arratia flow. We should note that models with finite speeds of particles moving under the influence of forces of mutual attraction were studied by Sinai (1992), Lifshits and Shi (2005), Giraud (2001, 2005), Bertoin (2002) and Vysotsky (2008a).

I.2. Description of the book

The book is divided into two volumes, each containing two parts. Part 1 of Volume 1 consists of basic concepts and methods developed for random evolutions. These methods are the elementary tools for the rest of the book, and they include many results in potential operators and the description of some techniques to find closed-form expressions in relevant applications.

Part 2 of Volume 1 comprises three chapters (3, 4 and 5) dealing with asymptotic results (Chapter 3) and applications ranging from random motion with different types of boundaries, reliability of storage systems, telegraph processes, an alternative formulation to the Black–Scholes formula in finance, fading evolutions, jump telegraph processes and estimation of the number of level crossings for telegraph processes (Chapters 4 and 5).

Part 1 of Volume 2 extends many of the results of the latter part of Volume 1 to higher dimensions and consists of two chapters (1 and 2). Chapter 1 has the importance of presenting novel results of the random motion of the realistic three-dimensional case that has barely been mentioned in the literature. Chapter 2 deals with the interaction of particles in Markov and semi-Markov media, a topic many researchers have a strong interest in.

Part 2 of Volume 2 discusses applications of Markov and semi-Markov motions in mathematical finance across three chapters (3, 4 and 5). It includes applications of the telegraph process in modeling a stock price dynamic (Chapter 3), pricing of variance, volatility, covariance and correlation swaps with Markov volatility (Chapter 4), and the same pricing swaps with semi-Markov volatilities (Chapter 5).

The following is a general overview of the chapters and sections of the book. Chapters 1 and 2 of Volume 1 review the literature on the topic of random evolutions and outline the main areas of research. Many of these auxiliary results are used throughout the book.

Section 1.1 outlines research directions on the theory of telegraph processes and their generalizations.

In section 1.2, we introduce the notion of the projector operator and the generalized inverse operator or potential for an invertible reduced operator used in perturbation theory for linear operators. In turn, this theory is often used in the study of the asymptotic distribution of probability for reaching a "hard to reach domain".

In section 1.3, we consider the notion of a semigroup of operators generated by a Markov process. We give the definitions of the infinitesimal operator, the stationary distribution and the potential of a Markov process. These concepts are used in Chapter 3 for the asymptotic analysis of large deviations of semi-Markov processes.

Section 1.4 provides a constructive definition of a semi-Markov process based on the concept of the Markov renewal process (MRP). The notion of the semi-Markov kernel, which is a key definition for MRP, is considered. For a semi-Markov process, we introduce some auxiliary processes, with which a semi-Markov process forms a two-component (or bivariate) Markov process, and for such a process the infinitesimal operator is presented.

In section 1.5, we consider the notion of a lumped Markov chain and describe a phase merging scheme.

Section 1.6 describes a stochastic switching process in Markov and semi-Markov environments. We define semigroup operators associated with this process and consider their infinitesimal operator. In addition, the concept of superposition of independent semi-Markov processes is considered.

In Chapter 2 of Volume 1 we introduce homogeneous random evolutions (HRE), the elementary definitions, classification and some examples. We also present the martingale characterization and an analogue of Dynkin's formula for HRE. Some other important topics covered in this chapter are limit theorems, weak convergence and diffusion approximations, which are useful for Part 2 of Volume 2.

In Chapter 3 of Volume 1 we consider the asymptotic distribution of a functional of the time for reaching "hard to reach" areas of the phase space by a semi-Markov process on the line.

Section 3.1 is devoted to the analysis of the asymptotic distribution of a functional related to the time to reach a level that is infinitely removed by a semi-Markov process on the set of natural numbers.

In section 3.2, we give asymptotic estimates for the distribution of residence times of the semi-Markov process in the set of states that expands when the condition of existence of the functional A is not fulfilled.

In section 3.3, we obtain the asymptotic expansion for the distribution of the first exit time from the extending subset of the phase space of the semi-Markov process embedded in the diffusion process.

In section 3.4, we obtain asymptotic expansions for the perturbed semigroups of operators of the respective three-variate Markov process (after the standard extension of the phase space of the perturbed random evolution $u_{\varepsilon}(t,x)$ in the semi-Markov media), provided that the evolution $u_{\varepsilon}(t,x)$ weakly converges to the diffusion process as $\varepsilon > 0$.

In section 3.5, we obtain asymptotic expansions under Kac's condition in the diffusion approximation for the distribution of a particle position, which performs a random walk in a multidimensional space with Markov switching.

Section 3.6 describes a novel financial formula as an alternative to the well-known Black–Scholes formula for modeling the dynamic behavior of stock markets. This new formula is based on the asymptotic expansion for the singularly perturbed random evolution in Markov media.

Chapter 4 of Volume 1 is devoted to the computation of the stationary distributions for random switched processes with reflecting boundaries in Markov and semi-Markov environments. These results are used for the calculation of the efficiency of inventory control systems with feedback, and for some reliability problems of systems that are modeled by Markov and semi-Markov evolution.

In section 4.1, we derive the stationary distribution of transport processes with delaying in the boundaries in Markov media. These results are applicable to the study of multiphase systems with several reservoirs.

Section 4.2 deals with the transport process with semi-Markov switching. We find the stationary measure for this process and it is described by the differential equation with phase space on the corresponding interval and constant vector field values that depend on the switching semi-Markov process with a finite set of states.

In section 4.3, we give examples of applications of this method for the calculation of the stationary distributions for random switched processes with reflecting boundaries. In particular, we compute the efficiency coefficient of a single and two-phase inventory control system with feedback.

In section 4.4, we apply random evolutions with delaying barriers to modeling the control of supply systems with feedback, by considering the semi-Markov switching process.

In Chapter 5 of Volume 1 we study various models of stochastic evolutions that generalize the Goldstein–Kac telegraph process and investigate their distributions.

Section 5.1 is devoted to one-dimensional semi-Markov evolutions in an Erlang environment. In section 5.1.1, we derive a hyperbolic differential equation of the telegraph type for the pdf of the position of a particle moving at finite velocity. The interarrival times between two successive changes of velocity have an Erlang distribution.

In section 5.1.2, a method of solution of this partial differential equation is developed by using monogenic functions associated with a finite-dimensional commutative algebra. Section 5.1.3 extends the results of section 5.1.2 to infinite-dimensional commutative algebras. In section 5.1.4, by using the methods from sections 5.1.2 and 5.1.3, we obtain the distribution of a one-dimensional random evolution in Erlang media.

In section 5.2, we find the distribution of limiting positions of a particle moving according to a fading evolution. We assume that the interarrival times are Erlang or uniform distributed for the switching process.

In section 5.3, differential and integral equations for jump random motions are introduced and some examples have been developed to illustrate the method.

Section 5.4 is devoted to the estimation of the number of level crossings by the telegraph process in Kac's condition.

In Chapter 1 of Volume 2 we study random motions in higher dimensions. In section 1.1, we study the random walk of a particle with constant absolute velocity, which changes its direction according to a uniform distribution on the unit sphere at the renewal epochs of a switching process.

Section 1.2 deals with random motion with uniformly distributed directions and random velocity for the motion of particles in one, two, three and four dimensions. Similar cases are considered in section 1.3 for the distribution of random motion at non-constant velocity in semi-Markov media, and in section 1.4 for Goldstein–Kac telegraph equations and random flights in higher dimensions, where the directions of the movement and the velocity change at the renewal epochs. The jump telegraph process in $\mathbb{R}^n$ is considered in section 1.5.

In Chapter 2 of Volume 2 we study a system of interacting particles with Markov and semi-Markov switching. Section 2.1 is devoted to an ideal gas model with finite velocity of molecules. In this section, we find the distribution of the first meeting time of two telegraph particles on the line, which started simultaneously from different points.

In section 2.1.2, we estimate the number of particle collisions. In section 2.1.4, we obtain an asymptotic estimate of the number of collisions of the system of telegraph particles when time goes to infinity. Unlike the previous section, the system has no boundaries and particles can move arbitrarily far away in a straight line. Section 2.2 is devoted to the generalization of results of section 2.1 to the case of semi-Markov switching processes. In section 2.2.1, we obtain a set of renewal-type equations for the Laplace transform of the first collision of two particles. In section 2.2.2, we consider an example regarding a particular case of semi-Markov switching of particle velocity, and we study the limiting properties of the distribution of the particle position. Section 2.2.3 is devoted to studying the case where the first time of collision of two particles has finite expectation. We should note that in all previous examples, including the Markov case, expectation of the first time of collision of two particles is infinite.

In Chapter 3 of Volume 2 the application of the telegraph process for option prices, as an alternative to the diffusion process in the Black-Scholes formula, is further studied through asymptotic estimation of the corresponding operators. Some numerical results and plots are also presented.

Pricing variance, volatility derivatives, covariance and correlation swaps for financial markets with Markov-modulated volatilities are the main topics of Chapter 4 of Volume 2. These results are extended to the case of semi-Markov modulated volatilities in Chapter 5 of Volume 2. Numerical results and some plots to illustrate these results are also presented in these last two chapters.

PART 1

Higher-dimensional Random Motions and Interactive Particles

1

Random Motions in Higher Dimensions

In this chapter, we discuss many of the previous results in further detail. Section 1.1 focuses on the study of an isotropic random motion at constant absolute velocity, where the direction alternations occur at renewal epochs of the Erlang or a general distribution. In recent years, the study of random walks with non-Markov switching processes is a topic of research. Many researchers have given special attention to the Pearson–Gamma random walk where the length of steps are Gamma distributed. In addition, the number of steps is related to the Dirichlet distribution. This type of motion is called a Pearson–Dirichlet motion. Apparently, the first study in this direction was done by Franceschetti (2007), in which the author studies a random walk in $\mathbb{R}^d$ ($d = 1, 2$) with uniformly distributed directions and the steps are i.i.d. random vectors of exponential length, and the walker travels a fixed total path. This means that the steps of the walk have joint Dirichlet distribution with all parameters equal to $q = 1$. By using the characteristic function technique, the author obtained a closed-form expression for the conditional distribution for the particle position after n steps, and he found a necessary condition for the uniform density of the particle position for a given number of steps in dimension d. It is mentioned in the paper that for dimensions $d \geq 3$ the random walk does not lead to a uniform density. Le Caër (2010, 2011) generalizes most of the results in Franceschetti (2007) to the case of multidimensional Pearson–Dirichlet random walk, whose parameters are all equal to $q > 0$. In Le Caër (2010), the author introduces and studies the so-called "Hyperspherical Uniform" (HU) Pearson–Dirichlet random walk. At renewal epochs (i.e. when the direction is changing), the HU random walk has a distribution in a unit disc identical to the distribution of the projection on the disc of a point uniformly distributed over the surface of a sphere of higher dimension. By using this HU property, Le Caër finds the pdf of the particle position that can be written in a closed-form expression with parameters $q = n - 1$, $n \geq 2$ and $q = n/2 - 1$, $n \geq 3$.

Another family of random motions is presented in Le Caër (2011), and it also has a closed-form expression for the distribution under the condition $q = n$. In this work, the author generalizes the approach developed in the paper by Beghin and Orsingher (2010a) for the parameter $q = 2$ on the plane, since he considers solutions for all integer $q = n \geq 2$.

In all of these works, the distribution of the particle position at renewal epochs of the switching directions process is studied. Basically, for the case of the non-Markov switching process, this task is reduced to studying the distribution of the non-Markov process through its embedded Markov chain. Here, we should mention also the paper by Letac and Piccioni (2014), in which the authors study Stiltjes transitions for a Pearson–Dirichlet walk, and by using these transitions they simplify the proof of similar results obtained by Le Caër, Orsingher and Beghin, as well as the limiting properties of the Dirichlet walk in the infinite growth of the number of steps.

We have contributed in high-dimensional random motions and these findings will be presented in this chapter. For instance, in section 1.1.1 we obtain formulas for the recursive calculations of the conditional characteristic functions of a random walk, where we are considering non-Markov switching processes with Erlang distribution for the interarrival times. Namely, we study changes of the conditional characteristic functions at each Poisson event (Pogorui and Rodríguez-Dagnino 2011).

In section 1.1.2, we derive the renewal equation for the characteristic function of the transition density for this multidimensional motion. Furthermore, by using this renewal equation we investigate some properties of the distribution function for the isotropic motion in two- and three-dimensional spaces. Section 1.2 is devoted to random motion with uniformly distributed directions and random velocity with examples for one-, two-, three- and four-dimensional cases. In section 1.2.1, the renewal equation for the characteristic function of isotropic motion with random velocity in a semi-Markov environment is obtained. Section 1.3 deals with multidimensional isotropic random motion at random speed, where the direction alternations occur according to the renewal epochs of a general distribution. Such a model represents a generalization of the model that only considers a constant velocity, which has been the typical assumption in most works related to this topic (Orsingher 1985; Orsingher and De Gregorio 2007; Orsingher and Ratanov 2007; Pinsky 1991; Franceschetti 2007; Le Caër 2010, 2011; Letac and Piccioni 2014; Pogorui and Rodríguez-Dagnino 2011; Pogorui 2011). We find the distribution at a fixed time of the random motions with some velocity distribution for one-, two-, three- and four-dimensional cases.

Section 1.3 deals with the characteristic function for the jump telegraph process in higher dimensions. In section 1.6.1, the explicit formula for this characteristic function is obtained. In section 1.4, we study the correspondence between a telegraph-type equation and random motion in three- and five-dimensional spaces.

1.1. Random motion at finite speed with semi-Markov switching directions process

In this section, we study random motions in $\mathbb{R}^n$ having uniformly distributed changes of direction at finite speed. The direction alternations occur according to a non-Markov switching process. In recent years, there have been many important works related to these processes (Beghin and Orsingher 2010a; Franceschetti 2007; Le Caër 2010, 2011; Letac and Piccioni 2014; Pogorui and Rodríguez-Dagnino 2011). In these studies, most of the authors are interested in non-Markov cases, where they can find closed-form expressions for the pdf and the conditional characteristic function of these semi-Markov transport processes.

In section 5.1 of volume 1 (Volume 1, Chapter 5), an isotropic random motion at finite speed with Erlang-k-distributed direction alternations is studied. Now, we will develop some relations for the recursive calculation of conditional characteristic functions of the random motion in the Erlang media considering that the switching process is not Markov. Namely, we study changes in the conditional characteristic functions at all Poisson epochs. In addition, we obtain the integral Volterra equation for the characteristic function of a Pearson–Gamma random walk with an integer parameter. This formulation aims to achieve a better understanding of the characteristic function of a random walk in the general semi-Markov medium as it satisfies a renewal integral equation.

In section 5.1 of volume 1 (Volume 1, Chapter 5), we also studied the case where the switching process is a general renewal process. In this section, we investigate some properties for the distribution function of the position of a particle in two- and three-dimensional space by using the integral equation for the characteristic functions of this distribution.

Let $\{\xi(t), t \geq 0\}$ be a renewal process such that $\xi(t) = \max\{m \geq 0 : \tau_m \leq t\}$, where $\tau_m = \sum_{k=0}^{m} \theta_k$, $\tau_0 = 0$ and $\theta_k \geq 0$, $k = 1, 2, \ldots$, are independent random variables denoting interarrival times. We assume that these random variables are identically distributed with cumulative distribution function (cdf) $G_\theta(t)$ and that there exists the pdf $g_\theta(t) = \frac{d}{dt}G(t)$. We will study the random motion of a particle that starts from the coordinate origin $\mathbf{0} = (0, 0, \ldots, 0)$ of the space $\mathbb{R}^n$, at time $t = 0$, and continues its motion with a constant velocity v along the direction $\boldsymbol{\eta}_0^{(n)}$, where $\boldsymbol{\eta}_0^{(n)} = (x_1, x_2, \ldots, x_n)$ is a random n-dimensional vector uniformly distributed on the sphere

$$\Omega_1^{n-1} = \{(x_1, x_2, \ldots, x_n) : x_1^2 + x_2^2 + \ldots + x_n^2 = 1\}.$$

At instant τ_1, the particle changes its direction to $\boldsymbol{\eta}_1^{(n)}$, where $\boldsymbol{\eta}_1^{(n)}$ and $\boldsymbol{\eta}_0^{(n)}$ are i.i.d. random vectors on Ω_1^{n-1}. Then, at instant τ_2 the particle changes its direction to $\boldsymbol{\eta}_2^{(n)}$, where $\boldsymbol{\eta}_2^{(n)}$, $\boldsymbol{\eta}_1^{(n)}$ and $\boldsymbol{\eta}_0^{(n)}$ are i.i.d. random vectors, and so on.

Denote by $\mathbf{x}^{(n)}(t)$, $t \geq 0$, the particle position at time t. We have that

$$\mathbf{x}^{(n)}(t) = v \sum_{j=1}^{\xi(t)} \boldsymbol{\eta}_{j-1}^{(n)} (\tau_j - \tau_{j-1}) + v \boldsymbol{\eta}_{\xi(t)}^{(n)} \left(t - \tau_{\xi(t)}\right).$$

Now let us denote as $\nu(t)$ the number of velocity alternations occurred in the interval $(0, t)$, and let us assume first a Poisson process for the epochs of these alternations. Then, the random variables θ_k are exponentially distributed with parameter λ, hence $g_\theta(t) = \lambda e^{-\lambda t} \mathbb{I}_{\{t \geq 0\}}$.

Furthermore, let us denote as $\mathbf{x}_m^{(n)}(t)$ the particle position at time $t > 0$ when $\nu(t) = m \geq 0$, i.e.

$$P\left(\mathbf{x}_m^{(n)}(t) \leq y\right) = P\left(\mathbf{x}^{(n)}(t) \leq y | \, \nu(t) = m\right).$$

It is not difficult to see that $\mathbf{x}_m^{(n)}(t)$ is given by

$$\mathbf{x}_m^{(n)}(t) = v \sum_{j=1}^{m+1} \boldsymbol{\eta}_{j-1}^{(n)} (\tau_j - \tau_{j-1}), \qquad [1.1]$$

where $\tau_{m+1} = t$.

The probabilistic properties of the random vector $\mathbf{x}_m^{(n)}(t)$ are completely determined by those of its projection $x_m^{(n)}(t) = v \sum_{j=1}^{m+1} \eta_{j-1}^{(n)} (\tau_j - \tau_{j-1})$ on a fixed line, where $\eta_j^{(n)}$ is the projection of $\boldsymbol{\eta}_j^{(n)}$ on the line.

Indeed, let us consider the conditional cdf

$$F_x(y | \, \nu(t) = m) = P\left(x_m^{(n)}(t) \leq y | \, \nu(t) = m\right).$$

Then, the conditional characteristic function (Fourier transform) $H_m(t, \alpha) = H_m(t)$ of $\mathbf{x}^{(n)}(t)$, where $\alpha = \|\boldsymbol{\alpha}\| = \sqrt{\alpha_1^2 + \alpha_2^2 + \ldots + \alpha_n^2}$, is given by

$$\begin{aligned} H_m(t) &= \mathbf{E}\left[\exp\left\{i\left(\boldsymbol{\alpha}, \mathbf{x}^{(n)}(t)\right)\right\} | \, \nu(t) = m\right] \\ &= \mathbf{E}\left[\exp\left\{i\left(\boldsymbol{\alpha}, \mathbf{x}_m^{(n)}(t)\right)\right\}\right] = \mathbf{E}\left[\exp\left\{i \|\boldsymbol{\alpha}\| \left(\mathbf{e}, \mathbf{x}_m^{(n)}(t)\right)\right\}\right] \\ &= \mathbf{E}\left[\exp\left\{i \|\boldsymbol{\alpha}\| x_m^{(n)}(t)\right\}\right] = \int_0^\infty \exp\{i \|\boldsymbol{\alpha}\| y\} dF_x(y | \, \nu(t) = m), \end{aligned}$$

where $x_m^{(n)}(t)$ is the projection of $\mathbf{x}_m^{(n)}(t)$ onto the unit vector $\mathbf{e} = \frac{\boldsymbol{\alpha}}{\|\boldsymbol{\alpha}\|}$ and it has a conditional cdf $F_x(y | \, \nu(t) = m)$.

Let denote by $f_{\eta^{(n)}}(x)$ the pdf of the projection $\eta_j^{(n)}$ of the vector $\boldsymbol{\eta}_j^{(n)}$ onto a fixed line. It is shown in Pogorui (2009b) that $f_{\eta^{(n)}}(x)$ is of the following form:

$$f_{\eta^{(n)}}(x) = \begin{cases} \frac{\Gamma\left(\frac{n}{2}\right)}{\sqrt{\pi}\Gamma\left(\frac{n-1}{2}\right)}\left(1-x^2\right)^{(n-3)/2} & , x \in [-1,1], \\ 0 & , x \notin [-1,1]. \end{cases} \qquad [1.2]$$

We can also consider the case $n = 1$ and it is easily seen that

$$f_{\eta^{(1)}}(x) = \frac{1}{2}\delta\left(|x| - 1\right).$$

It is also straightforward to verify that

$$H_m(t) = \mathbf{E}\left[\exp\left\{i\left\|\boldsymbol{\alpha}\right\| v \sum_{j=1}^{m+1} \eta_{j-1}^{(n)}\left(\tau_j - \tau_{j-1}\right)\right\}\right]$$

$$= \frac{m!}{t^m}\int_0^t ds_{\text{-}} \int_{s_1}^t ds_2 \ldots \int_{s_{m-1}}^t ds_m \mathbf{E}\left[\exp\left\{i\left\|\boldsymbol{\alpha}\right\| v \sum_{j=1}^{m+1} \eta_{j-1}^{(n)}\left(s_j - s_{j-1}\right)\right\}\right]$$

$$= \frac{m!}{t^m}\int_0^t ds_1 \int_{s_1}^t ds_2 \ldots \int_{s_{m-1}}^t ds_m \left\{\prod_{j=1}^{m+1} \varphi_{\eta^{(n)}}\left(\left\|\boldsymbol{\alpha}\right\| v\left(s_j - s_{j-1}\right)\right)\right\},$$

where $s_0 = 0$, $s_{m+1} = t$ and $\varphi_{\eta^{(n)}}(\lambda) = \mathbf{E}\left[e^{-i\lambda\eta_j^{(n)}}\right] = \int_{-\infty}^{\infty} e^{-i\lambda x} f_{\eta^{(n)}}(x)dx$ is the characteristic function of $\eta_j^{(n)}$. It is well known that (Pogorui and Rodríguez-Dagnino 2011)

$$\varphi(t) = 2^{\frac{n-2}{2}}\Gamma\left(\frac{n}{2}\right)\frac{J_{\frac{n-2}{2}}(\alpha t v)}{(\alpha t v)^{\frac{n-2}{2}}}. \qquad [1.3]$$

1.1.1. *Erlang-K-distributed direction alternations*

Let us assume a fixed integer $K \geq 2$. Now, suppose that the particle changes its direction just at epochs τ_{Kj}, $j = 1, 2, \ldots$, i.e.

$$\mathbf{x}_m^{(n)}(t) = v\sum_{j=1}^{\lfloor \frac{m}{K} \rfloor} \boldsymbol{\eta}_{K(j-1)}^{(n)}\left(\tau_{Kj} - \tau_{K(j-1)}\right)$$

$$+ v\,\boldsymbol{\eta}_{K\lfloor \frac{m}{K} \rfloor}^{(n)}\left(\tau_{m+1} - \tau_{K\lfloor \frac{m}{K} \rfloor}\right), \qquad [1.4]$$

where $\tau_{m+1} = t$ and $\sum_{j=1}^{0} \boldsymbol{\eta}_{K(j-1)}^{(n)}\left(\tau_{Kj} - \tau_{K(j-1)}\right) = 0$.

Hence, defining $c = \alpha v$, we have for $m = lK + r$, $l = 0, 1, 2, \ldots, 0 \leq r < K$

$$H_l^{(r)}(t) = H_m(t) = \mathbf{E}\left[\exp\left\{ic\mathbf{x}_m^{(n)}(t)\right\}\right]$$

$$= \frac{(lK+r)!}{t^{lK+r}} \int_0^t ds_1 \int_{s_1}^t ds_2 \ldots \int_{s_{lK+r-1}}^t ds_{lK+r}$$

$$\times \left\{\prod_{j=1}^{l} \varphi\left((s_{Kj} - s_{Kj-K})\right)\varphi\left((t - s_{Kl})\right)\right\},$$

where $\prod_{j=1}^{0} \varphi\left((s_{Kj} - s_{K(j-1)})\right) = 1$.

We should note that $H_0^{(r)}(t) = \varphi(t)$.

The random variables $\tau_{Kj} - \tau_{K(j-1)}$, $j \geq 1$, are Erlang-K distributed, and we may define the renewal process $\xi_K(t) = \max\left\{j \geq 0 : \tau_j^{(K)} \leq t\right\}$, $t \geq 0$, where $\tau_j^{(K)} = \sum_{l=0}^{j} \theta_l^{(K)}$, $\tau_0^{(K)} = 0$ and $\theta_l^{(K)}$, $l = 1, 2, \ldots$, are i.i.d. interarrival times having an Erlang-K pdf $g_\theta(t) = \frac{\lambda^K t^{K-1}}{(K-1)!} e^{-\lambda t} I_{\{t \geq 0\}}$.

Thus, the stochastic process $\mathbf{x}^{(n)}(t)$ can be considered as the random evolution of the particle in the non-Markov Erlang-K medium $\xi_K(t)$.

The functions $H_l^{(r)}(t)$, $l = 0, 1, 2, \ldots$ are the conditional characteristic functions of $\mathbf{x}^{(n)}(t)$. We should remember that $\nu(t) = m$ represents m Poisson events occurred in $[0, t)$, and the number of direction alternations $\nu_K(t)$ of $\mathbf{x}^{(n)}(t)$ is given by $\nu_K(t) = \left\lfloor \frac{\nu(t)}{K} \right\rfloor$.

For $l = 0, 1, 2, \ldots; 0 \leq r < K$, let us consider the following multiple integral:

$$I_l^{(r)}(t) = \int_0^t ds_1 \int_{s_1}^t ds_2 \ldots \int_{s_{lK+r-1}}^t ds_{lK+r}$$

$$\times \left\{\prod_{j=1}^{l} \varphi\left((s_{Kj} - s_{Kj-K})\right)\varphi\left((t - s_{Kl})\right)\right\}, \qquad [1.5]$$

where $s_0 = 0$. Let us state the notation $\psi_K(t) = \dfrac{t^{K-1}}{(K-1)!}\varphi(t)$.

THEOREM 1.1.– For any $l \geq 0$, we have the following recursive relation:

$$I_{l+1}^{(r)}(t) = \int_0^t \frac{u^{K-1}}{(K-1)!}\varphi((u))\, I_l^{(r)}(t-u)\, du = \psi_K * I_l^{(r)}(t), \qquad [1.6]$$

where we have by assumption that $I_0^{(r)}(t) = \frac{t^r}{r!}\varphi(t)$, $r = 0, 1, \ldots, K-1$.

PROOF.– Let us prove [1.6] by induction arguments. First, suppose that $l = r = 0$, then

$$I_1^{(0)}(t) = \int_0^t ds_1 \int_{s_1}^t ds_2 \ldots \int_{s_{K-1}}^t ds_K \left\{\varphi((s_K))\, I_0^{(0)}(c(t-s_K))\right\}$$

$$= \left[s_1 \int_{s_1}^t ds_2 \ldots \int_{s_{K-1}}^t ds_K \left\{\varphi((s_K))\, I_0^{(0)}(c(t-s_K))\right\}\right]_0^t$$

$$+ \int_0^t s_1 \int_{s_1}^t ds_3 \ldots \int_{s_{K-1}}^t ds_K \left\{\varphi((s_K))\, I_0^{(0)}(c(t-s_K))\right\} ds_1$$

$$= \int_0^t \frac{s_1^2}{2} \int_{s_1}^t ds_4 \ldots \int_{s_{K-1}}^t ds_K \left\{\varphi((s_K))\, I_0^{(0)}(c(t-s_K))\right\} ds_1$$

$$\vdots$$

$$= \int_0^t \frac{s_1^{K-1}}{(K-1)!}\varphi((s_1))\, I_0^{(0)}(c(t-s_1))\, ds_1 = \psi_K * I_0^{(0)}(t). \qquad [1.7]$$

Now, suppose that [1.6] is valid for any $l \leq N-2$, for a fixed $N > 2$ and $r = 0$.

The integral $I_{l+1}^{(0)}(t)$ for $l = N-1$ can be represented in the following form:

$$I_N = \int_0^t ds_1 \ldots \int_{s_{K-1}}^t ds_K \left\{\varphi(s_K) \int_{s_K}^t ds_{K+1}\right.$$

$$\ldots \int_{s_{2K-1}}^t ds_{2K} \left\{\varphi((s_{2K}-s_K)) \int_{s_{2K}}^t ds_{2K+1} \ldots \left\{\int_{s_{K(N-1)-1}}^t ds_{K(N-1)}\right.\right.$$

$$\left.\left.\left.\ldots \int_{s_{KN-1}}^t ds_{KN}\varphi((s_{KN}-s_{K(N-1)}))\, \varphi((t-s_{kN}))\right\}\ldots\right\}. \qquad [1.8]$$

Let us consider the following interior integral in $I_N^0(t)$ (with respect to $s_{K(N-1)+1}$) in [1.8]. Let us make the change of variables

$\xi_j = s_{K(N-1)+j} - s_{K(N-1)}$, $j = 1, 2, \ldots, K$, and by taking into account [1.7], we obtain

$$\int_{s_{K(N-1)}}^{t} ds_{K(N-1)+1} \cdots \int_{s_{KN-1}}^{t} ds_{KN}\varphi\left(\left(s_{KN} - s_{K(N-1)}\right)\right)\varphi\left(\left(t - s_{KN}\right)\right)$$

$$= \int_{0}^{t-s_{K(N-1)}} d\xi_1 \int_{\xi_1}^{t-s_{K(N-1)}} \cdots \int_{\xi_{K-1}}^{t-s_{K(N-1)}} d\xi_K\varphi\left(\xi_K\right)\varphi\left(\left(t - s_{K(N-1)} - \xi_K\right)\right)$$

$$= \psi_K * I_1^{(0)}\left(t - s_{K(N-1)}\right).$$

Now, let us deal with the next interior integral in [1.8] (with respect to $s_{K(N-2)+1}$). By defining the change of variables $\xi_j = s_{K(N-2)+j} - s_{K(N-2)}$, $j = 1, 2, \ldots, K$, we obtain

$$\int_{s_{K(N-2)}}^{t} ds_{K(N-2)+1} \cdots \int_{s_{K(N-1)-1}}^{t} ds_{K(N-1)}\varphi\left(\left(s_{K(N-1)} - s_{K(N-2)}\right)\right)$$

$$\times I_1^{(0)}\left(t - s_{K(N-1)}\right)$$

$$= \int_{0}^{t-s_{K(N-1)}} d\xi_1 \int_{\xi_1}^{t-s_{K(N-1)}} \cdots \int_{\xi_{K-1}}^{t-s_{K(N-1)}} d\xi_K\varphi\left(\xi_K\right)$$

$$\times I_1^{(0)}\left(t - s_{K(N-2)} - \xi_K\right) = \psi_K * I_1^{(0)}\left(\left(t - s_{K(N-2)}\right)\right).$$

This is in accordance with [1.7] and the induction assumption.

By continuing this procedure, we can obtain, after the $(N-1)$th step,

$$I_N^{(0)}(t) = \psi_K * I_{N-1}^{(0)}(t).$$

Next, it can be easily verified that for any $0 \leq r \leq K-1$

$$\int_{s_K}^{t} ds_{K+1} \cdots \int_{s_{K+r-1}}^{t} ds_{K+r} = \frac{\left(t - s_K\right)^r}{r!}.$$

Hence, for all $0 < r \leq K-1$, we have

$$I_1^{(r)}(t) = \int_0^t ds_1 \int_{s_1}^t ds_2 \ldots \int_{s_{K-1}}^t ds_K \left\{\varphi\left(\left(s_K\right)\right)\varphi\left(\left(t - s_K\right)\right)\right\} \int_{s_K}^t ds_{K+1}$$

$$\cdots \int_{s_{K+r-1}}^{t} ds_{K+r}$$

$$= \int_0^t ds_1 \int_{s_1}^t ds_3 \cdots \int_{s_{K-1}}^t ds_K \left\{ \varphi\left((s_K)\right) \frac{(t-s_K)^r}{r!} \varphi\left((t-s_K)\right) \right\}$$

$$= \int_0^t \frac{s_1^{K-1}}{(K-1)!} \varphi\left((s_1)\right) I_0^{(r)}\left(c\left(t-s_K\right)\right) ds_1 = \psi_K * I_0^{(r)}(t)$$

and by using the recursive relation [1.6] it can be proved for $0 < r \leq K-1$, in a similar manner as we did for the case $r = 0$.

COROLLARY 1.1.– It follows from [1.7] and [1.8] that

$$I_l^{(r)}(t) = \psi_K^{l*} * I_0^{(r)}(t).$$

where $l*$ is the l-fold convolution, $l > 0$, $I_0^{(r)}(t) = \frac{t^r}{r!}\varphi(t)$, $r = 0, 1, \ldots, K-1$.

COROLLARY 1.2.– For any $l \geq 0$, the conditional characteristic functions satisfy the following equations:

$$H_{l+1}^{(r)}(t) = \frac{(K(l+1)+r)!}{t^{K(l+1)+r}(Kl+r)!} \int_0^t u^{Kl+r} \psi_K(u) H_l^{(r)}(t-u)\, du.$$

PROOF.– From [1.6], it follows that

$$H_{l+1}^{(r)}(t) = \frac{(K(l+1)+r)!}{t^{K(l+1)+r}} I_{l+1}^{(r)}(t)$$

$$= \frac{(K(l+1)+r)!}{t^{K(l+1)+r}(Kl+r)!} \int_0^t u^{Kl+r} \psi_K(u) \left[\frac{(Kl+r)!}{u^{Kl+r}} I_l^{(r)}(t-u) \right] du$$

$$= \frac{(K(l-1)+r)!}{t^{K(l+1)+r}(Kl+r)!} \int_0^t u^{Kl+r} \psi_K(u) H_l^{(r)}(t-u)\, du.$$

From equation [1.11] and using $H_0^{(r)}(t) = \varphi(t)$, we can calculate $H_l^{(r)}(t)$ for all $l \geq 1$, $r = 0, 1, \ldots, K-1$.

Many authors (such as Le Caër 2010, 2011; Beghin and Orsingher 2010a; De Gregorio and Orsingher 2012; and De Gregorio 2014), have studied the conditional distributions of the particle position at renewal epochs with velocities switched by a renewal process with Dirichlet-distributed interarrival times. These authors studied the characteristic functions $H_l^{(0)}(t)$, $l \geq 1$ in terms of the conditional characteristic function $H_l^{(r)}(t)$.

THEOREM 1.2.– The characteristic function $H(t) = \mathbf{E}\left[\exp\left\{ic\mathbf{x}^{(n)}(t)\right\}\right]$, $t \geq 0$, of the random motion $\mathbf{x}^{(n)}(t)$ is a solution of the following renewal equation:

$$H(t) = e^{-\lambda t} \sum_{r=0}^{K-1} \frac{t^r}{r!}\varphi(t) + \lambda^K \int_0^t \psi_K (t-u) e^{-\lambda(t-u)} H(u)\, du, \qquad [1.9]$$

or in the convolution form

$$H(t) = \varphi(t)\left(1 - F_K(t)\right) + \left(\varphi f_K\right) * H(u),$$

where $F_K(t)$ is Erlang-K distributed and $f_K(t) = \frac{d}{dt} F_K(t)$.

PROOF.– It is easily seen that

$$H(t) = \exp\left\{i\left(\boldsymbol{\alpha}, \mathbf{x}^{(n)}(t)\right)\right\} = \sum_{l=0}^{+\infty}\sum_{r=0}^{K-1} H_l^{(r)}(t) P\left(\nu(t) = Kl + r\right)$$

$$= e^{-\lambda t} \sum_{l=0}^{\infty}\sum_{r=0}^{K-1} \lambda^{Kl+r} I_l^{(r)}(t).$$

Now, we need to show that for any $t > 0$ the series $\sum_{l=0}^{\infty}\sum_{r=0}^{K-1} \lambda^{Kl+r} I_l^{(r)}(t)$ uniformly converges with respect to $c = \alpha v$.

Indeed, taking into account that $|\varphi(\lambda)| = \left|\mathbf{E}\left[e^{-i\lambda\eta_j^{(n)}}\right]\right| \leq 1$, we have

$$\left|I_l^{(r)}(t)\right| \leq \int_0^t ds_1 \int_{s_1}^t ds_2 \ldots \int_{s_{lK+r-1}}^t ds_{lK+r} = \frac{t^{lK+r}}{(lK+r)!}.$$

By using this bound, we state

$$\left|\sum_{l=0}^{\infty}\sum_{r=0}^{K-1} \lambda^{Kl+r} I_l^{(r)}(t)\right| \leq \sum_{l=0}^{\infty}\sum_{r=0}^{K-1} \frac{t^{lK+r}}{(lK+r)!} = e^{\lambda t}.$$

Therefore, $H(t)$ can be represented by the uniformly convergent series as follows:

$$H(t) = e^{-\lambda t} \sum_{l=0}^{\infty}\sum_{r=0}^{K-1} \lambda^{Kl+r} I_l^{(r)}(t) = e^{-\lambda t} \sum_{l=0}^{\infty}\sum_{r=0}^{K-1} \psi_K^{l*} * I_0^{(r)}(t). \qquad [1.10]$$

Hence, we have

$$H(t) = \exp\left\{i\left(\boldsymbol{\alpha}, \mathbf{x}^{(n)}(t)\right)\right\} = \sum_{l=0}^{+\infty}\sum_{r=0}^{K-1} H_l^{(r)}(t) P\left(\nu(t) = Kl + r\right)$$

$$= e^{-\lambda t}\left\{\sum_{r=0}^{K-1}\frac{t^r}{r!}\varphi(t)+\int_0^t \psi_K(t-u)\left(\sum_{l=1}^{\infty}\sum_{r=0}^{K-1}\lambda^{Kl+r}I_{l-1}^{(r)}(u)\right)du\right\}$$

$$= e^{-\lambda t}\left\{\sum_{r=0}^{K-1}\frac{t^r}{r!}\varphi(t)+\lambda^K\int_0^t \psi_K(t-u)\left(\sum_{l=0}^{\infty}\sum_{r=0}^{K-1}\lambda^{Kl+r}I_{l}^{(r)}(u)\right)du\right\}$$

$$= e^{-\lambda t}\left\{\sum_{r=0}^{K-1}\frac{t^r}{r!}\varphi(t)+\lambda^K\int_0^t \psi_K(t-u)\,e^{\lambda u}H(u)\,du\right\}.$$

In the following sections, equation [1.9] is generalized for an arbitrary distribution of time between consecutive changes of the particle direction and for random absolute velocity of the particle.

EXAMPLE 1.1.– Let us consider a three-dimensional random motion with Erlang-2 distribution for times between consecutive changes of the particle direction. In this case, $f_{\eta^{(3)}}(x) = \frac{1}{2}I\{-1 \le x \le 1\}$. Therefore, $\varphi(t) = \mathbf{E}\left[e^{ict\eta^{(3)}}\right] = \frac{\sin ct}{ct}$, and $H_0^{(0)}(t) = \varphi(\alpha vt) = \frac{\sin \alpha vt}{\alpha vt} = H_0^{(1)}(t)$, where $\alpha = \sqrt{\alpha_1^2+\alpha_2^2+\alpha_3^2}$.

It is well known that the function $4\pi R\dfrac{\sin \alpha R}{\alpha}$ is the Fourier transform for the function of simple layer $\delta_{S(R)}$ (Vladimirov 1996).

Denote by $\overline{H}_l(t) = H_l^{(0)}(t)P[\nu(t)=2l] + H_l^{(1)}(t)P[\nu(t)=2l+1]$, the characteristic function of the particle position $\mathbf{x}_l^{(3)}(t)$ at time t provided that l changes of velocity are occurred. Hence,

$$P\left(\mathbf{x}^{(3)}(t)\in d\mathbf{x}/\nu_2(t)=0\right) = P\left(\mathbf{x}_0^{(3)}(t)\in d\mathbf{x}\right)$$

$$= \mathcal{F}^{-1}\left(\overline{H}_0(t)\right)dx_1dx_2dx_3 = \frac{\delta_{S(vt)}}{4\pi(vt)^2}\left(e^{-\lambda t}+\lambda te^{-\lambda t}\right)dx_1dx_2dx_3,$$

where $\mathbf{x} = (x_1, x_2, x_3)$.

Let us calculate $H_1^{(0)}(t)$ and $H_1^{(1)}(t)$. It is easily verified that for the three-dimensional space, we have

$$H_1^{(0)}(t) = \frac{2}{t^2}\int_0^t \psi(t-u)H_0^{(0)}(u)\,du$$

$$= \frac{2}{\alpha^2t^2v^2}\int_0^t \frac{(t-u)\sin(\alpha v(t-u))}{(t-u)}\frac{\sin(\alpha vu)}{u}du$$

$$= \frac{2}{\alpha^2t^2v^2}\int_0^t \frac{\sin(\alpha v(t-u))\sin(\alpha vu)}{u}du$$

$$
\begin{aligned}
&= \frac{1}{\alpha^2 t^2 v^2} \left(\mathrm{Si}\,(2\alpha vt) \sin(\alpha vt) + \mathrm{Ci}\,(2\alpha vt) \cos(\alpha vt)\right. \\
&\quad \left. - \left(\ln(2\alpha vt) + \gamma\right) \cos(\alpha vt)\right) \\
&= \frac{1}{\alpha^2 t^2 v^2} \left(\mathrm{Si}\,(2\alpha vt) \sin(\alpha vt) + \int_0^{2\alpha vt} \frac{\cos \xi - 1}{\xi} d\xi \cos(\alpha vt) \right),
\end{aligned}
$$

where γ is Euler's constant.

Denote by $f(\mathbf{x} \mid \nu(t) = 2)\, d\mathbf{x} = P\left(\mathbf{x}^{(3)}(t) \in d\mathbf{x} \mid \nu(t) = 2\right)$ the conditional pdf of a particle position assuming that there was only one change of speed occurred and after this no more Poisson event occurred.

Then we have $f(\mathbf{x} \mid \nu(t) = 2) = \mathcal{F}^{-1}\left(H_1^{(0)}(t)\right)$, where $\mathcal{F}^{-1}$ is the inverse Fourier transform.

Hence,

$$
f(\mathbf{x} \mid \nu(t) = 2) = \frac{1}{4\pi \|\mathbf{x}\| (vt)^2} \ln\left(\frac{vt + \|\mathbf{x}\|}{vt - \|\mathbf{x}\|}\right) \to \infty, \quad \|\mathbf{x}\| \uparrow vt.
$$

Consequently, for the continuous part $f_c(\mathbf{x})$ of the distribution of the particle position, we obtain

$$
f(\mathbf{x} \mid \nu(t) = 2) \le f_c(\mathbf{x}) \to \infty, \quad \|\mathbf{x}\| \uparrow vt.
$$

Thus, in this case, the explosion effect holds in a similar manner as the case of the exponential switching process.

For the conditional pdf $f(\mathbf{x} \mid \nu(t) = 3)$ of the particle position assuming that only one change of speed occurred and after that exactly one Poisson event occurred, we have $f(\mathbf{x} \mid \nu(t) = 3) = \mathcal{F}^{-1}\left(H_1^{(1)}(t)\right)$. For $H_1^{(1)}(t)$, we obtain

$$
\begin{aligned}
H_1^{(1)}(t) &= \frac{6}{\alpha^2 t^3 v^2} \int_0^t u \frac{(t-u) \sin(\alpha v (t-u))}{(t-u)} \frac{\sin(\alpha v u)}{u} du \\
&= \frac{6}{t^3 v^2 \alpha^2} \int_0^t \sin(\alpha v (t-u)) \sin(\alpha v u) du = 3 \frac{\sin(\alpha vt) - \alpha vt \cos(\alpha vt)}{t^3 v^3 \alpha^3} \\
&= \chi(\alpha).
\end{aligned}
$$

Since the function $\chi(\alpha)$ depends only on $\alpha = \sqrt{\alpha_1^2 + \alpha_2^2 + \alpha_3^2}$, the function $f(\mathbf{x} \mid \nu(t) = 3)$ depends only on $\rho = \|\mathbf{x}\| = \sqrt{x_1^2 + x_2^2 + x_3^2}$, that is $f(\mathbf{x} \mid \nu(t) = 3) = \phi_1^{(1)}(\rho)$, $\rho \le vt$.

Thus, we have

$$\chi(\alpha) = \frac{(2\pi)^{3/2}}{\sqrt{\alpha}} \int_0^{vt} \phi_1^{(1)}(\rho)\,\rho^{3/2} J_{1/2}(\rho\alpha)d\rho$$

$$= \frac{4\pi}{\alpha} \int_0^{vt} \phi_1^{(1)}(\rho)\,\rho \sin(\rho\alpha)d\rho.$$

It is easily verified that if $\phi_1^{(1)}(\rho) = \frac{3}{4\pi t^3 v^3} I_{\{\rho \leq vt\}}$, then

$$\frac{4\pi}{\alpha} \int_0^{vt} \frac{3}{4\pi t^3 v^3} \rho \sin \alpha\rho d\rho = \chi(\alpha).$$

This means that $f(\mathbf{x} \,|\, \nu(t) = 3) = \frac{3}{4\pi t^3 v^3}\mathbb{I}_{\{\|\mathbf{x}\| \leq vt\}}$ is the inverse three-dimensional Fourier transform of the function $\chi(\alpha)$.

Therefore, after exactly three Poisson events the particle position is uniformly distributed over the unit ball $B(vt) = \{\|\mathbf{x}\| < vt\}$ as follows:

$$f(\mathbf{x} \,|\, \nu(t) = 3) = \frac{3}{4\pi t^2 v^3}\mathbb{I}_{\{\|\mathbf{x}\| \leq vt\}}.$$

1.1.2. *Some properties of the random walk in a semi-Markov environment and its characteristic function*

Now, we consider the general cdf $G(t)$, which has the pdf $g(t)$. By using renewal theory one can easily prove that the characteristic function $H(t)$ is a solution of the following Volterra renewal-type integral equation:

$$H(t) = (1 - G(t))\,\varphi(t) + \int_0^t g(u)\,\varphi(u)\,H(t-u)\,du. \qquad [1.11]$$

We will generalize equation [1.11] in the following lemma to the case where the absolute velocity of the particle is a random variable. We should recall that equation [1.11] has a unique solution in the case of a continuous kernel $g\,\varphi$, since it is a Volterra equation.

LEMMA 1.1.– Suppose that $g(t) > 0$ for all $t \geq 0$. Then for $n = 2, 3$, we have

$$f_n(t, \mathbf{x}) \uparrow \infty \text{ as } \|\mathbf{x}\| \uparrow vt.$$

PROOF.– Since $f_n(t, \mathbf{x}) = \mathcal{F}^{-1}(H(t))$, where $\mathcal{F}^{-1}$ is the n-dimensional inverse Fourier transform of $H(t) = (H(t, \|\alpha\|))$ with respect to $\boldsymbol{\alpha}$, then it follows from [1.11] that

$$f_n(t, \mathbf{x}) = \mathcal{F}^{-1}(H(t)) = (1 - G(t))\,\frac{\Gamma\left(\frac{n}{2}\right)\delta\left(v^2t^2 - \|\mathbf{x}\|^2\right)}{2\pi^{\frac{n}{2}}(vt)^{n-1}}$$

$$+\frac{\left(\Gamma\left(\frac{n}{2}\right)\right)^2}{4\pi^n}\int_{\|\mathbf{u}\|\le vt}\int_0^t \frac{\left(1-G\left(t-s\right)\right)\delta\left(v^2(t-s)^2-\|\mathbf{x}-\mathbf{u}\|^2\right)}{\left(v\left(t-s\right)\right)^{n-1}}$$

$$\times\frac{g\left(s\right)\delta\left(v^2s^2-\|\mathbf{u}\|^2\right)}{\left(vs\right)^{n-1}}dsd\mathbf{u}+\ldots$$

For $n=3$, we have $\varphi(t)=\dfrac{\sin(vt\,\|\boldsymbol{\alpha}\|)}{vt\,\|\boldsymbol{\alpha}\|}$.

It is well known that

$$\mathcal{L}\left(\frac{\sin(vt\,\|\boldsymbol{\alpha}\|)}{vt\,\|\boldsymbol{\alpha}\|}\right)=\frac{1}{v\,\|\boldsymbol{\alpha}\|}\operatorname{arctg}\left(\frac{v\,\|\boldsymbol{\alpha}\|}{s}\right).$$

Thus, we have

$$\frac{\left(\Gamma\left(\frac{3}{2}\right)\right)^2}{4\pi^3}\int_{\|\mathbf{u}\|\le vt}\int_0^t\frac{\delta\left(v^2(t-s)^2-\|\mathbf{x}-\mathbf{u}\|^2\right)}{\left(v\left(t-s\right)\right)^2}\frac{\delta\left(v^2s^2-\|\mathbf{u}\|^2\right)}{\left(vs\right)^2}dsd\mathbf{u}$$

$$=\mathcal{F}^{-1}\left(\frac{1}{v^2\|\boldsymbol{\alpha}\|^2}\mathcal{L}^{-1}\left[\left(\operatorname{arctg}\left(\frac{v\,\|\boldsymbol{\alpha}\|}{s}\right)\right)^2\right]\right)=\frac{1}{4\pi v^2t^2\,\|\mathbf{x}\|}\ln\left(\frac{vt+\|\mathbf{x}\|}{vt-\|\mathbf{x}\|}\right).$$

Since for any $s\ge 0$, the function $g\left(s\right)>0$, we have

$$C_t=\inf_{0\le s\le t}\left(1-G\left(t-s\right)\right)g\left(s\right)>0$$

and

$$\frac{C_t}{4\pi v^2t^2\,\|\mathbf{x}\|}\ln\left(\frac{vt+\|\mathbf{x}\|}{vt-\|\mathbf{x}\|}\right)\le f_3\left(t,\mathbf{x}\right).$$

Hence, $f_3\left(t,\mathbf{x}\right)\uparrow\infty$ as $\|\mathbf{x}\|\uparrow vt$, i.e. we have the explosive effect.

For the case $n=2$ (i.e. $\mathbf{u}$, $\mathbf{x}\in\mathbb{R}^2$), we have

$$\frac{2}{4\pi^2}\int_0^t\int_{\|\mathbf{u}\|\le vt}\frac{\delta\left(v^2(t-s)^2-\|\mathbf{x}-\mathbf{u}\|^2\right)}{\left(v\left(t-s\right)\right)^2}\frac{\delta\left(v^2s^2-\|\mathbf{u}\|^2\right)}{\left(vs\right)^2}dsd\mathbf{u}$$

$$=\frac{\left(v^2s^2-\|\mathbf{u}\|^2\right)^{-\frac{1}{2}}}{4\pi vt}.$$

By following the same approach, we can show that $f_2\left(t,\mathbf{x}\right)\uparrow\infty$ as $\|\mathbf{x}\|\uparrow vt$.

In the case of an Erlang pdf $g(t)$, the explosive effect also holds because in these spaces of dimension $n=2,3$ there are not enough directions to leave the "layer" around the sphere of singularity.

1.2. Random motion with uniformly distributed directions and random velocity

Now, we deal with uniformly distributed direction of motion or isotropic motion at random speed or velocity where the direction alternations occur according to the renewal epochs of a general distribution. We derive the renewal equation for the characteristic function of the transition density of the multidimensional motion. Then, by using the renewal equation, we study the behavior of the transition density near the sphere of its singularity for one-, two-, three- and four-dimensional cases. To illustrate our solution methodology, we present detailed calculations of many solvable examples.

1.2.1. *Renewal equation for the characteristic function of isotropic motion with random velocity in a semi-Markov media*

Most papers concerning isotropic random motion on the multidimensional space are devoted to the analysis of models in which motions are driven by the homogeneous Poisson process. Another typical assumption is to consider constant speed or constant velocity; their models are basically Markov processes (e.g. (Orsingher and De Gregorio 2007; Orsingher and Beghin 2006; Pinsky 1991; Pogorui and Rodríguez-Dagnino 2011; Pogorui 2011) and references therein). In this section, we consider multidimensional random motions with uniformly distributed directions having random velocity. We will find a renewal equation for the characteristic function of the distribution of the particle position of this motion.

For the case of four-dimensional space, we give an example having an explosive effect, i.e. the pdf of the particle position goes to infinity when approaching the sphere of singularity if the absolute speed of the particle is some random variable.

As mentioned before, we consider the renewal process $\nu(t) = \max\{m \geq 0 : \ \tau_m \leq t\}, t \geq 0$, where $\tau_m = \sum_{k=0}^{m} \theta_k$, $\tau_0 = 0$ and $\theta_k \geq 0$, $k = 1, 2, \ldots$, are i.i.d. random variables denoting the interarrival times. Let $G(t)$ be the cdf of θ_1 and we assume that there exists the pdf $g(t) = \frac{d}{dt}G(t)$.

We will study the random motion of a particle that starts from the coordinate origin $\mathbf{0} = (0, 0, \ldots, 0)$ of the space $\mathbb{R}^n$ at time $t = 0$ and continues its motion with a velocity $\gamma_0 > 0$ along the direction $\boldsymbol{\eta}_0^{(n)}$, where γ_0 is a random variable with the cdf $Z(v)$ and there exists the pdf $z(v) = \frac{d}{dv}Z(v)$, $v \geq 0$, and the direction $\boldsymbol{\eta}_0^{(n)} = (x_1, x_2, \cdots, x_n)$ is a random n-dimensional vector uniformly distributed on the sphere

$$\Omega_1^{n-1} = \left\{(x_1, x_2, \cdots, x_n) : x_1^2 + x_2^2 + \ldots + x_n^2 = 1\right\}.$$

The random variables γ_0 and $\boldsymbol{\eta}_0^{(n)}$ are independent.

Then, the particle moves in straight lines and it changes its direction at renewal epochs τ_m, $m = 1, 2, \ldots$, much in the same manner, which was considered in Volume 1, Chapter 5, with the only difference that at epoch τ_k the particle changes not only the direction of its movement from $\boldsymbol{\eta}_{k-1}^{(n)}$ to $\boldsymbol{\eta}_k^{(n)}$ but with the direction it also changes the absolute velocity from γ_{k-1} to γ_k, where γ_{k-1} and γ_k are i.i.d. with the distribution $Z(v)$.

Denote by $\mathbf{x}^{(n)}(t), t \geq 0$, the particle position at time t. As in the previous section, we have

$$\mathbf{x}^{(n)}(t) = \sum_{j=1}^{\nu(t)} \boldsymbol{\eta}_{j-1}^{(n)} \gamma_{j-1} \left(\tau_j - \tau_{j-1}\right) + \boldsymbol{\eta}_{\nu(t)}^{(n)} \gamma_{\nu(t)} \left(t - \tau_{\nu(t)}\right). \qquad [1.12]$$

Let us consider the projection $x^{(n)}(t)$ of $\mathbf{x}^{(n)}(t)$ onto a fixed line. Then we have

$$x^{(n)}(t) = \sum_{j=1}^{m+1} \eta_{j-1}^{(n)} \gamma_{j-1} \left(\tau_j - \tau_{j-1}\right) + \eta_{\nu(t)}^{(n)} \gamma_{\nu(t)} \left(t - \tau_{\xi(t)}\right)$$

where $\eta_j^{(n)}$ is the projection of $\boldsymbol{\eta}_j^{(n)}$ onto the line.

Denote as $F_x(y) = P\left(x_m^{(n)}(t) \leq y\right)$. For the characteristic function of $\mathbf{x}^{(n)}(t)$, we have

$$H(t) = \mathbf{E}\left[\exp\left\{i\left(\boldsymbol{\alpha}, \mathbf{x}^{(n)}(t)\right)\right\}\right] = \mathbf{E}\left[\exp\left\{i\left\|\boldsymbol{\alpha}\right\|\left(\mathbf{e}, \mathbf{x}^{(n)}(t)\right)\right\}\right]$$

$$= \mathbf{E}\left[\exp\left\{i\left\|\boldsymbol{\alpha}\right\| x^{(n)}(t)\right\}\right] = \int_0^\infty \exp\left\{i\left\|\boldsymbol{\alpha}\right\| y\right\} dF_x(y),$$

where $\|\boldsymbol{\alpha}\| = \sqrt{\alpha_1^2 + \alpha_2^2 + \cdots + \alpha_n^2}$.

Thus, the probabilistic properties of $\mathbf{x}^{(n)}(t)$ are completely defined by the properties of $x^{(n)}(t)$.

Denote by $f_{\eta^{(n)}}(x)$ the pdf of $\eta_j^{(n)}$. From [1.2], it follows that

$$f_{\eta^{(n)}}(x) = \begin{cases} \frac{\Gamma\left(\frac{n}{2}\right)}{\sqrt{\pi}\Gamma\left(\frac{n-1}{2}\right)} \left(1 - x^2\right)^{(n-3)/2} & , x \in [-1, 1], \\ 0 & , x \notin [-1, 1]. \end{cases} \qquad [1.13]$$

Hence, it is easily verified that $F_{\eta^{(n)}v}(x) = P\left(\eta_j^{(n)}\gamma_j \le x\right)$ is of the following form:

$$F_{\eta^{(n)}\gamma}(x) = \begin{cases} \frac{1}{2} + \frac{\Gamma(\frac{n}{2})}{\sqrt{\pi}\Gamma(\frac{n-1}{2})} \int_0^1 Z\left(\frac{x}{t}\right)\left(1-t^2\right)^{(n-3)/2} dt, & x \ge 0, \\ \frac{1}{2} - \frac{\Gamma(\frac{n}{2})}{\sqrt{\pi}\Gamma(\frac{n-1}{2})} \int_0^1 Z\left(-\frac{x}{t}\right)\left(1-t^2\right)^{(n-3)/2} dt, & x < 0. \end{cases} \quad [1.14]$$

Let us denote by $\varphi(t) = \mathbf{E}\left[e^{it\alpha\gamma_j\eta_j^{(n)}}\right] = \int_{-\infty}^{\infty} e^{it\alpha x} dF_{\eta^{(n)}\gamma}(x)$ the characteristic function of the random variable $\eta_j^{(n)}\gamma_j$. As it is mentioned in section 1.5.1., if $\gamma_j = v = const > 0$, then the function $\varphi(t)$ is as follows:

$$\varphi(t) = 2^{\frac{n-2}{2}}\Gamma\left(\frac{n}{2}\right)\frac{J_{\frac{n-2}{2}}(\alpha t v)}{(\alpha t v)^{\frac{n-2}{2}}},$$

where $\alpha = \|\boldsymbol{\alpha}\|$.

In the case where γ_j is a random variable with pdf $z(v)$, $v \ge 0$, it is easy to verify that $\varphi(t)$ is of the following form:

$$\varphi(t) = \mathbf{E}\left[e^{it\alpha\gamma_j\eta_j^{(n)}}\right] = 2^{\frac{n-2}{2}}\Gamma\left(\frac{n}{2}\right)\int_0^\infty \frac{J_{\frac{n-2}{2}}(\alpha t v)}{(\alpha t v)^{\frac{n-2}{2}}} z(v)\, dv. \quad [1.15]$$

THEOREM 1.3.– The characteristic function $H(t), t \ge 0$, is a solution of the following Volterra renewal-type integral equation:

$$H(t) = (1 - G(t))\,\varphi(t) + \int_0^t g(u)\,\varphi(u)\,H(t-u)\,du. \quad [1.16]$$

PROOF.– From equation [1.12], it follows that

$$H(t) = \mathbf{E}\left[\exp\left\{i\left(\boldsymbol{\alpha}, \mathbf{x}^{(n)}(t)\right)\right\}\right]$$

$$= \mathbf{E}\left[\exp\left\{i\left(\boldsymbol{\alpha}, \sum_{j=1}^{\xi(t)} \boldsymbol{\eta}_{j-1}^{(n)}\gamma_{j-1}\theta_j + \gamma_{\xi(t)}\boldsymbol{\eta}_{\xi(t)}^{(n)}\left(t - \tau_{\xi(t)}\right)\right)\right\}\right]$$

$$= (1 - G(t))\, Ee^{it\gamma_0\left(\boldsymbol{\alpha},\boldsymbol{\eta}_0^{(n)}\right)} + \int_0^t g(u)\, Ee^{iu\gamma_0\left(\boldsymbol{\alpha},\boldsymbol{\eta}_0^{(n)}\right)} H(t-u)\,du.$$

To complete the proof, we mention that $\varphi(t) = \mathbf{E}\left[e^{it\gamma_0\left(\boldsymbol{\alpha},\boldsymbol{\eta}_0^{(n)}\right)}\right]$.

Now let us denote by $f_n(t, \mathbf{x})$ the pdf of the particle position at time t. Therefore, $f_n(t, \mathbf{x}) = \mathcal{F}^{-1}(H(t))$, where $\mathcal{F}^{-1}$ is the n-dimensional inverse Fourier transform with respect to $\boldsymbol{\alpha}$.

Let us study functions $f_n(t, \mathbf{x})$ for $n = 1, 2, 3, 4$.

1.2.2. *One-dimensional case*

Let us consider the one-dmensional case, i.e. $n = 1$, with constant velocity $v > 0$, $\varphi(t) = \cos(\alpha t v)$. We have, for the exponential pdf $g(t) = \lambda e^{-\lambda t}\mathbb{I}_{\{t\geq 0\}}$, the renewal equation

$$H(t) = e^{-\lambda t}\cos(\alpha t v) + \lambda \int_0^t e^{-\lambda u}\cos(\alpha u v) H(t-u)\, du. \qquad [1.17]$$

We should note that the process $\mathbf{x}^{(1)}(t)$ is not the telegraph process because at renewal epochs the particle may not change its direction.

Denote by $h(t) = H(t)e^{\lambda t}$, then from equation [1.17] it follows that

$$h(t) = \cos(\alpha t v) + \lambda \int_0^t \cos(\alpha u v)\, h(t-u)\, du.$$

Since $\mathcal{L}(\cos(\alpha t v)) = \int_0^\infty e^{-su}\cos(\alpha u v) du = \frac{s}{s^2+v^2\alpha^2}$, then after applying the Laplace transform to [1.17] (or its equivalent in $h(t)$) we obtain

$$\mathcal{L}(h(t)) = \int_0^\infty e^{-st}h(t)dt = \frac{s}{s^2 - \lambda s + v^2\alpha^2}.$$

Thus, the inverse Laplace transform gives

$$H(t) = e^{-\lambda t}h(t) = e^{-\frac{\lambda t}{2}}\left(\cosh\left(\frac{t}{2}\sqrt{\lambda^2 - 4v^2\alpha^2}\right)\right.$$
$$\left. + \frac{\lambda \sinh\left(\frac{t}{2}\sqrt{\lambda^2 - 4v^2\alpha^2}\right)}{\sqrt{\lambda^2 - 4v^2\alpha^2}}\right). \qquad [1.18]$$

We should recall the formula from Bateman (1954, p. 57, no. 47).

$$\frac{\sinh\left(t\sqrt{1-\alpha^2}\right)}{\sqrt{1-\alpha^2}} = \frac{1}{2}\int_{-t}^{t} I_0\left(\sqrt{t^2 - x^2}\right) e^{ix\alpha}dx.$$

This can be tailored to our case as

$$\frac{\lambda \sinh\left(\frac{t}{2}\sqrt{\lambda^2 - 4v^2\alpha^2}\right)}{\sqrt{\lambda^2 - 4v^2\alpha^2}} = \frac{\lambda}{2v}\int_{-tv}^{tv} I_0\left(\frac{\lambda}{2v}\sqrt{(tv)^2 - x^2}\right) e^{ix\alpha}dx.$$

Therefore, by applying the inverse Fourier transform (with respect to α) to equation [1.18], we obtain

$$f_1(t,x) = \frac{1}{2}\delta\left((vt)^2 - x^2\right)e^{-\frac{\lambda t}{2}}$$

$$+\frac{e^{-\frac{\lambda t}{2}}}{2v}\left[\frac{\lambda}{2}I_0\left(\frac{\lambda}{2v}\sqrt{(tv)^2 - x^2}\right) + \frac{\partial}{\partial t}I_0\left(\frac{\lambda}{2v}\sqrt{(tv)^2 - x^2}\right)\right]$$

$$= \frac{1}{2}\delta\left((vt)^2 - x^2\right)e^{-\frac{\lambda t}{2}} + \frac{\lambda}{4v}I_0\left(\frac{\lambda}{2v}\sqrt{(tv)^2 - x^2}\right)e^{-\frac{\lambda t}{2}}$$

$$+\frac{\lambda t}{4}\frac{I_1\left(\sqrt{(tv)^2 - x^2}\right)}{\sqrt{(tv)^2 - x^2}}e^{-\frac{\lambda t}{2}}.$$

Thus, $f_1(t,x)$ is a solution of the following telegraph equation:

$$\frac{\partial^2 f}{\partial t^2} + \lambda\frac{\partial f}{\partial t} - v^2\frac{\partial^2 f}{\partial x^2} = 0, \quad f(x,0) = \delta(x),$$

which defines the process with the exponential distribution with parameter $\lambda/2$ for interarrival times of the switching directions process. This is because the switching process for this random walk is a Markov process with the phase space $\{0,1\}$ and the following infinitesimal matrix:

$$Q = \begin{pmatrix} -\lambda/2 & \lambda/2 \\ \lambda/2 & -\lambda/2 \end{pmatrix}.$$

We should recall that for the Goldstein–Kac process the infinitesimal matrix is as follows:

$$2Q = \begin{pmatrix} -\lambda & \lambda \\ \lambda & -\lambda \end{pmatrix}.$$

Therefore, if we consider only real change of direction, the previous process is a Goldstein–Kac process with the parameter $\lambda/2$.

Suppose that γ_1 is a random variable with pdf

$$z(v) = \frac{2}{\pi\sqrt{1-v^2}}, \quad 0 \le v \le 1.$$

This pdf is the density of distribution of the projection of a vector uniformly distributed on a unit semicircle onto a line that passes through the center of the circle (see section 1.4.2). Then, we obtain

$$\varphi(t) = \int_0^1 \cos(\alpha t v)\frac{2}{\pi\sqrt{1-v^2}}dv = J_0(\alpha t). \qquad [1.19]$$

For $g(t) = \lambda e^{-\lambda t} I_{\{t \geq 0\}}$, we have

$$H(t) = e^{-\lambda t} J_0(\alpha t) + \lambda \int_0^t e^{-\lambda(t-u)} J_0(\alpha(t-u)) H(u) du. \quad [1.20]$$

Let us state as $h(t) = e^{\lambda t} H(t)$. Then equation [1.20] is of the form:

$$h(t) = J_0(\alpha t) + \lambda \int_0^t J_0(\alpha(t-u)) h(u) du. \quad [1.21]$$

It is well known (Bateman 1954) that the Laplace transform of the Bessel function $J_0(\alpha t)$ with respect to t is as follows:

$$\mathcal{L}(J_0(\alpha t)) = \frac{1}{\sqrt{s^2 + \alpha^2}}.$$

By obtaining the Laplace transform in equation [1.21], we have

$$\mathcal{L}(h(t)) = \frac{1}{\sqrt{s^2 + \alpha^2}} + \frac{\lambda}{\sqrt{s^2 + \alpha^2}} \mathcal{L}(h(t)).$$

Hence,

$$\mathcal{L}(h(t)) = \frac{1}{\sqrt{s^2 + \alpha^2} - \lambda} = \frac{1}{\lambda} \sum_{n=1}^{\infty} \left(\frac{\lambda}{\sqrt{s^2 + \alpha^2}} \right)^n.$$

The inverse Laplace transform of the function $\frac{1}{\lambda}\left(\frac{\lambda}{\sqrt{s^2+\alpha^2}}\right)^n$ with respect to s is of the following form:

$$\mathcal{L}^{-1}\left(\frac{1}{\lambda}\left(\frac{\lambda}{\sqrt{s^2+\alpha^2}}\right)^n\right) = \frac{\sqrt{\pi}}{\Gamma\left(\frac{n}{2}\right)} \left(\frac{\lambda^2 t}{2}\right)^{\frac{n-1}{2}} \frac{J_{\frac{n-1}{2}}(\alpha t)}{\alpha^{\frac{n-1}{2}}}.$$

Therefore,

$$H(t) = e^{-\lambda t} \sum_{n=1}^{\infty} \frac{\sqrt{\pi}}{\Gamma\left(\frac{n}{2}\right)} \left(\frac{\lambda^2 t}{2}\right)^{\frac{n-1}{2}} \frac{J_{\frac{n-1}{2}}(\alpha t)}{\alpha^{\frac{n-1}{2}}}.$$

Then, by reducing the inverse Fourier transform to the Hankel transform, we obtain

$$f_1(t,x) = \mathcal{F}^{-1}(H(t)) = e^{-\lambda t} \sum_{n=1}^{\infty} \frac{\lambda^{n-1}}{\left(\Gamma\left(\frac{n}{2}\right)\right)^2 2^{n-1}} \frac{\left(t^2 - x^2\right)^{\frac{n-2}{2}}}{(2t)^{\frac{n-1}{2}}} \mathbb{I}_{\{t>|x|\}}.$$

For the first term, $\mathcal{F}^{-1}\left(e^{-\lambda t} J_0(\alpha t)\right) = \frac{e^{-\lambda t}}{\pi\sqrt{t^2-x^2}} \mathbb{I}_{\{t>|x|\}}$. Hence, it is easily seen that the pdf $f_1(t,x)$ satisfies $f_1(t,x) = \mathcal{F}^{-1}(H(t)) \to +\infty, \ |x| \uparrow t$.

Therefore, we have an explosive effect at lines $|x| = t$ while for the constant velocity v the explosive effect does not appear.

1.2.3. *Two-dimensional case*

We should note that in case of constant velocity $\gamma_j = v > 0$, we have $\varphi(t) = J_0(\alpha t v)$.

Now, let us suppose that the velocity γ_j is a random variable with the following heavy-tail pdf:

$$z(v) = \frac{v}{(1+v^2)^{\frac{3}{2}}}, \quad v \geq 0.$$

Then,

$$\varphi(t) = \int_0^\infty J_0(\alpha t v)\, z(v) dv = \int_0^\infty \frac{J_0(\alpha t v)\, v}{(1+v^2)^{\frac{3}{2}}} dv = e^{-t\alpha}.$$

Now, let us assume an exponential pdf $g(t) = \lambda e^{-\lambda t} \mathbb{I}_{\{t \geq 0\}}$; hence, we have the renewal equation

$$H(t) = e^{-\lambda t} e^{-\alpha t} + \lambda \int_0^t e^{-\lambda(t-u)} e^{-\alpha(t-u)} H(u)\, du. \quad [1.22]$$

By solving equation [1.22], we have

$$H(t) = e^{-\alpha t}.$$

Now, we can obtain the two-dimensional inverse Fourier transform of $H(t) = H(t, \alpha)$ with respect to α, and we can apply the Hankel transform (Bochner and Chandrasekhar 1949) to obtain the pdf of $\mathbf{x}^{(2)}(t)$

$$f_2(t, x) = \frac{1}{2\pi} \int_0^\infty e^{-\alpha t} \alpha J_0(\alpha x)\, d\alpha = \frac{t}{2\pi (t^2 + x^2)^{\frac{3}{2}}},$$

$$x = |\mathbf{x}|, \quad \mathbf{x} = (x_1, x_2).$$

1.2.4. *Three-dimensional case*

1.2.4.1. *Isotropic transport process with random velocity in* $\mathbb{R}^3$

The isotropic random motion of a particle at constant velocity in $\mathbb{R}^n$ driven by a homogeneous Poisson process has been intensively studied by many researchers. For planar ($n = 2$) random motion, the explicit form of the pdf for the particle position at a fixed time was obtained by several methods (Stadje 2007; Masoliver *et al.* 1993b). However, for the most important three-dimensional case, the pdf for the continuous part of the distribution of the process was given in the form a complicated integral

(Stadje 2007), which can hardly be evaluated precisely. We remove the condition of constant velocity of the particle, and consider the case of a random speed, but distributed in the way such that it is close to a constant speed. We obtain the closed form for the pdf of the particle position in $\mathbb{R}^3$.

Let $\{\nu(t)\}$ be a Poisson process with rate $\lambda > 0$ starting at $\nu(0) = 0$. We define the time of the kth jump by $\tau_k = \inf\{t : \nu(t) = k\}$, $k \geq 0$. The random variable τ_k is also called the Poisson event and $\theta_k = \tau_{k+1} - \tau_k$ is called the kth waiting time (or interarrival time).

We will study the random motion of a particle that starts its motion from the origin $\mathbf{0} = (0, 0, 0)$ of the space $\mathbb{R}^3$, at time $t = 0$, and it continues its motion with a velocity $v_0 > 0$ along a direction $\boldsymbol{\eta}_0$ in $\mathbb{R}^3$. At instant τ_1, the particle changes its direction to $\boldsymbol{\eta}_1$ and continues its motion with a velocity $v_1 > 0$ along $\boldsymbol{\eta}_1$. At instant τ_2, the particle changes its direction to $\boldsymbol{\eta}_2$ and continues its motion with a velocity $v_2 > 0$ along $\boldsymbol{\eta}_2$, and so on. We assume that $\{v_i, i = 0, 1, 2, \ldots\}$ is a sequence of i.i.d. positive random variables with cdf $Z(s)$ and pdf $z(s) = dZ(s)/ds$, $s \geq 0$, and $\boldsymbol{\eta}_i$, $i = 0, 1, 2, \ldots$ are i.i.d. random three-dimensional vectors uniformly distributed on the unit sphere $\Omega_1^2 = \left\{(x_1, x_2, x_3) : x_1^2 + x_2^2 + x_3^2 = 1\right\}$. The random variables v_i are independent of the random variables $\boldsymbol{\eta}_j$, $i, j \in \{0, 1, 2, \ldots\}$.

Denote by $\mathbf{x}(t)$, $t \geq 0$ the particle position at time t. We have that

$$\mathbf{x}(t) = \sum_{j=1}^{\nu(t)} \boldsymbol{\eta}_{j-1} v_{j-1} \theta_{j-1} + \boldsymbol{\eta}_{\nu(t)} v_{\nu(t)} \left(t - \tau_{\nu(t)}\right). \qquad [1.23]$$

As we know (Pogorui and Rodríguez-Dagnino 2012), the characteristic function $H(t) = \mathbf{E}\left[\exp\{i(\boldsymbol{\alpha}, \mathbf{x}(t))\}\right]$ satisfies the following renewal equation:

$$H(t) = e^{-\lambda t}\varphi(t) + \lambda \int_0^t e^{-\lambda u}\varphi(u) H(t-u)\, du, \qquad [1.24]$$

where

$$\varphi(t) = \int_0^\infty z(v) \frac{\sin(\alpha t v)}{\alpha t v} dv, \qquad \alpha = |\boldsymbol{\alpha}|.$$

As far as we know, in the three-dimensional case for a constant velocity of a particle, the distribution of the particle position has not been obtained before, and this case is relevant for applications. Now, we consider the case of random velocity where the particle velocity has the following pdf:

$$z(s) = \frac{4s^2}{\pi\sqrt{1-s^2}}, \quad 0 \leq s < 1. \qquad [1.25]$$

This velocity is "almost equal" to $v = 1$. By making this assumption, we are able to obtain the distribution of particle position in the closed form.

$$\varphi(t) = \int_0^{\infty} \frac{4v^2}{\pi\sqrt{1-v^2}} \frac{\sin(\alpha t v)}{\alpha t v} dv = \frac{2J_1(at)}{at}.$$

Hence, we have the following equation:

$$H(t) = \varphi(t)e^{-\lambda t} + \lambda \int_0^t \varphi(u) e^{-\lambda u} H(t-u)\, du.$$

This renewal equation [1.24] has a unique solution of the following form Kovalenko *et al.* (1983):

$$H(t) = e^{-\lambda t} \sum_{n=0}^{\infty} \lambda^n [\varphi(t)]^{*(n+1)}$$

The relation for the Bessel function (Prudnikov *et al.* 1986, p.190, formula 2.12.33.8), is well known.

$$\int_0^t \frac{J_m(a(t-u))}{t-u} \frac{J_n(au)}{u} du = \frac{m+n}{mn} \frac{J_{n+m}(at)}{t}. \qquad [1.26]$$

By using equation [1.26], we obtain

$$[\varphi(t)]^{*(k)} = \frac{(k-1)\,2^{k-1}}{a^{k-1}} \int_0^t \frac{J_{k-1}(a(t-u))}{(t-u)} \frac{2J_1(au)}{au} du = k\, 2^k \frac{J_k(at)}{a^k t},$$

$k = 2, 3, \ldots$

where $[\varphi(t)]^{*(1)} = \varphi(t)$.

It is easily verified that

$$[\varphi(t)]^{*(k)} = k\, 2^k \frac{J_k(at)}{a^k t}.$$

The inverse Fourier (Hankel) transform of $[\varphi(t)]^{*(k)}$ is as follows:

$$I_k = \mathcal{F}^{-1}\left([\varphi(t)]^{*(k)}, \alpha\right) = \begin{cases} \frac{k2^k\sqrt{2/\pi}}{(k+1)!t^{k+1}} \sqrt{(t^2-x^2)^{2k-3}} & , k \text{ is even} \\ \frac{k2^k\sqrt{2/\pi}}{k!t^{k+1}} \sqrt{(t^2-x^2)^{2k-3}} & , k \text{ is odd,} \end{cases}$$

where $x = \|\mathbf{x}\|$.

Therefore, the pdf $f(t, \mathbf{x})$ of the particle position at time t is as follows:

$$f(t,x) = \mathcal{F}^{-1}(H(t))$$
$$= e^{-\lambda t}\sqrt{2/\pi}\left(e^{-\lambda t}\sqrt{2/\pi}\left(\sum_{m=1}^{\infty}\frac{(2m-1)\ 2^{(2m-1)}\lambda^{(2m-1)}}{(2m-1)!\,t^{2m}}\right.\right.$$
$$\left.\left.\times\sqrt{(t^2-x^2)^{4m-5}}+\sum_{l=1}^{\infty}\frac{2l\ 4^l\lambda^{2l}}{(2l+1)!\,t^{2l+1}}\sqrt{(t^2-x^2)^{4l-3}}\right)\right). \qquad [1.27]$$

The following formula is well known (Arfken *et al.* 2012, pp. 544–545 and 547–548):

$$\Gamma\left(n+\frac{1}{2}\right)=\frac{(2n-1)!}{2^n}\sqrt{\pi}. \qquad [1.28]$$

By using [1.28] and calculating the sum of series, we get for $x^2 < t^2$

$$f(t,x) = \mathcal{F}^{-1}(H(t)) = e^{-\lambda t}\sqrt{2}\left(\sum_{l=1}^{\infty}\frac{l\ 2^l\lambda^{2l}}{\Gamma(l+3/2)\,t^{2l+1}}\sqrt{(t^2-x^2)^{4l-3}}\right.$$
$$\left.+\sum_{m=1}^{\infty}\frac{(2m-1)\,2^{(m-1)}\lambda^{(2m-1)}}{\Gamma(m+1/2)\,t^{2m}}\sqrt{(t^2-x^2)^{4m-5}}\right)$$
$$=\frac{e^{-\lambda t}\sqrt{2}}{4\sqrt{\pi}\lambda t^2\Lambda^3}\left[4\lambda t\Lambda^{3/2}+8\lambda^2\Lambda^{5/2}+\left(\lambda^2\sqrt{128\Lambda^5}-t^2\sqrt{8\Lambda}\right.\right.$$
$$\left.\left.+\frac{16\lambda^3\sqrt{2\Lambda^7}}{t}\right)e^{\frac{2\lambda^2\Lambda^2}{t^2}}\int_0^{\sqrt{2}\lambda\Lambda/t}e^{-y^2}dy\right],$$

where $\Lambda = t^2 - x^2$.

REMARK 1.1.– *It is well known that random walks are used to model polymers (Banchoff 1976). One important characteristic of a polymer is the distribution of the distance between starting and ending points of its molecules and the result obtained may be interesting from this point of view.*

1.2.4.2. *Isotropic transport process with random velocity in $\mathbb{R}^3$: small parameter approximation*

Now, we will study a random motion in $\mathbb{R}^3$ with random velocity, distributed according to Maxwell's law of distribution of velocities. We believe that this model is a natural alternative to the Wiener model of Brownian motion. In fact, this is a more realistic model because of the two following facts: first, the velocity of the particles in this model has the Maxwell distribution, unlike the Wiener process, where the absolute particle velocity is infinite. Second, unlike the Wiener particle path, the particle in this model has a non-zero free path.

Let us consider a random motion $x(t)$, which is determined by equation [1.23], where the random velocity v_i is distributed according to the Maxwell distribution with pdf

$$z\left(v\right)=\sqrt{\frac{2}{\pi}}v^2\,e^{-\frac{1}{2}v^2},\quad v>0.$$

Then, it follows that

$$\varphi(t)=\sqrt{2/\pi}\int_0^\infty v^2\,e^{-\frac{1}{2}v^2}\frac{\sin\left(\alpha tv\right)}{\alpha tv}dv=e^{-\frac{t^2\alpha^2}{2}}.$$

Hence, the equation for the characteristic function of the distribution of particle position is as follows:

$$H(t)=e^{-\lambda t}e^{-\frac{t^2\alpha^2}{2}}+\lambda\int_0^t e^{-\lambda u}e^{-\frac{u^2\alpha^2}{2}}H\left(t-u\right)du.$$

Denote as $h(t)=e^{\lambda t}H(t)$, then we have

$$h(t)=e^{-\frac{1}{2}t^2\alpha^2}+\lambda\int_0^t e^{-\frac{1}{2}u^2\alpha^2}h\left(t-u\right)du.$$

Since the parameter $\varepsilon=1/\lambda$ is very small (e.g. in most gas models), we will study the perturbed integral equation

$$\varepsilon h(t)=\varepsilon e^{-\frac{1}{2}t^2\alpha^2}+\int_0^t e^{-\frac{1}{2}u^2\alpha^2}h\left(t-u\right)du.$$

A solution of such equations can be found by using the following expansion (Angell and Olmstead 1987; Shubin 2006):

$$h(t)=\sum_{k=0}^\infty\varepsilon^k\left(u_k(t)+w_k\left(\tau\right)\right),\tag{1.29}$$

where $\tau=t/\varepsilon$, the terms $u_k(t)$ are called regular, and $w_k\left(\tau\right)$ are called singular terms of [1.29].

For $u_0(t)$, we have

$$\varepsilon u_0(t)=\varepsilon e^{-\frac{1}{2}t^2\alpha^2}+\int_0^t e^{-\frac{1}{2}u^2\alpha^2}u_0\left(t-u\right)du.\tag{1.30}$$

From [1.30], it follows

$$\int_0^t e^{-\frac{1}{2}u^2\alpha^2} u_0\,(t-u)\,du = 0. \qquad [1.31]$$

It is well known that equation [1.31] has only the trivial solution

$$u_0(t) = 0.$$

For $w_0\,(\tau)$, we have

$$\varepsilon w_0\,(\tau) = \varepsilon e^{-\frac{1}{2}t^2\alpha^2} + \int_0^t e^{-\frac{1}{2}(t-u)^2\alpha^2} w_0\left(\frac{u}{\varepsilon}\right) du.$$

By letting $s = \frac{u}{\varepsilon}$, we have

$$w_0\,(\tau) = e^{-\frac{1}{2}\tau^2\alpha^2\varepsilon^2} + \int_0^\tau e^{-\frac{1}{2}(\tau-s)^2\alpha^2\varepsilon^2} w_0\,(s)\,ds. \qquad [1.32]$$

In equation [1.32], we can only use the first term of the expansion for the exponent, so

$$e^{-\frac{1}{2}\tau^2\alpha^2\varepsilon^2} = 1 - \frac{1}{2}\tau^2\alpha^2\varepsilon^2 + \frac{1}{4}\tau^4\alpha^4\varepsilon^4 + \cdots, \qquad [1.33]$$

$$e^{-\frac{1}{2}(\tau-s)^2\alpha^2\varepsilon^2} = 1 - \frac{1}{2}(\tau-s)^2\alpha^2\varepsilon^2 + \frac{1}{4}(\tau-s)^4\alpha^4\varepsilon^4 + \cdots.$$

Hence, we obtain

$$w_0\,(\tau) = 1 + \int_0^\tau w_0\,(s)\,ds.$$

It is easily verified that

$$w_0\,(\tau) = e^{\tau}.$$

Next, for $u_1(t)$ we have

$$\varepsilon^2 u_1(t) = \varepsilon e^{-\frac{1}{2}t^2\alpha^2} + \varepsilon \int_0^t e^{-\frac{1}{2}(t-u)^2\alpha^2} u_1\,(u)\,du$$

or

$$e^{-\frac{1}{2}t^2\alpha^2} = \int_0^t e^{-\frac{1}{2}(t-u)^2\alpha^2} u_1\,(u)\,du.$$

For $w_1(\tau)$, we have

$$\varepsilon^2 w_1(\tau) = \varepsilon\left(e^{-\frac{1}{2}t^2\alpha^2} - 1\right) + \varepsilon \int_0^t e^{-\frac{1}{2}(t-u)^2\alpha^2} w_1\left(\frac{u}{\varepsilon}\right) du.$$

Hence, by changing variable $s = \frac{u}{\varepsilon}$, we have

$$w_1(\tau) = -\varepsilon \frac{1}{2}\tau^2\alpha^2 + \int_0^\tau e^{-\frac{1}{2}(\tau-s)^2\alpha^2\varepsilon^2} w_1(s)\, du$$

or

$$w_1(\tau) = \int_0^\tau e^{-\frac{1}{2}(\tau-s)^2\alpha^2\varepsilon^2} w_1(s)\, du$$

and as a consequence $w_1(\tau) = 0$.

For $w_2(\tau)$

$$\varepsilon^3 w_2(\tau) = \varepsilon\left(e^{-\frac{1}{2}t^2\alpha^2} - 1\right) + \varepsilon^2 \int_0^t e^{-\frac{1}{2}(t-u)^2\alpha^2} w_1\left(\frac{u}{\varepsilon}\right) du$$

$$+ \int_0^t \left(e^{-\frac{1}{2}(t-u)^2\alpha^2} - 1\right) w_0\left(\frac{u}{\varepsilon}\right) du.$$

Introducing the change of variable $s = \frac{u}{\varepsilon}$, we have

$$\varepsilon^3 w_2(\tau) = \varepsilon\left(e^{-\frac{1}{2}t^2\alpha^2} - 1\right) + \varepsilon^3 \int_0^\tau e^{-\frac{1}{2}(t-u)^2\alpha^2} w_2(s)\, ds$$

$$+ \int_0^\tau \left(e^{-\frac{1}{2}(\tau-s)^2\alpha^2\varepsilon^2} - 1\right) w_0(s)\, ds.$$

Whence, we have

$$w_2(\tau) = -\frac{1}{2}\tau^2\alpha^2 + \int_0^\tau w_2(s)\, ds - \frac{1}{2}\alpha^2 \int_0^\tau (\tau - s)^2 w_0(s)\, ds.$$

Taking into account that $w_0(\tau) = e^\tau$, we have

$$w_2(\tau) = -\frac{1}{2}\tau^2\alpha^2 + \int_0^\tau w_2(s)\, ds - \frac{1}{2}\alpha^2\left(2e^\tau - \tau^2 - 2\tau - 1\right)$$

$$w_2(\tau) = \int_0^\tau w_2(s)\, ds - \alpha^2 e^\tau + \alpha^2\tau + \frac{1}{2}\alpha^2$$

$$w_2(0) = \frac{1}{2}\alpha^2$$

$$\frac{d}{d\tau}w_2(\tau) = w_2(\tau) - \alpha^2 e^{\tau} + \alpha^2$$

$$w_2(\tau) = -\alpha^2 - \alpha^2 \tau e^{\tau} + Ce^{\tau}$$

$$C = \frac{1}{2}\alpha^2$$

$$w_2(\tau) = \frac{1}{2}\alpha^2 e^{\tau} - \alpha^2 - \alpha^2 \tau e^{\tau}.$$

$$\varepsilon\varepsilon^4 w_4(\tau) = \varepsilon\varepsilon^4 \frac{1}{4}\tau^4\alpha^4 + \varepsilon^4 \int_0^t e^{-\frac{1}{2}(t-u)^2\alpha^2} w_4\left(\frac{u}{\varepsilon}\right) du$$

$$+\varepsilon^2 \int_0^t e^{-\frac{1}{2}(t-u)^2\alpha^2} w_2\left(\frac{u}{\varepsilon}\right) du + \int_0^t e^{-\frac{1}{2}(t-u)^2\alpha^2} w_0\left(\frac{u}{\varepsilon}\right) du$$

$$w_4(\tau) = \frac{1}{4}\tau^4\alpha^4 + \int_0^{\tau} w_4(s)\, ds - \int_0^{\tau} \frac{(\tau-s)^2\alpha^2}{2} w_2(s)\, ds$$

$$+\int_0^t \frac{(\tau-s)^4\alpha^4}{4} w_0(s)\, ds.$$

Hence,

$$\frac{d}{d\tau}w_4(\tau) = w_4(\tau) + \frac{\alpha^4}{12}\left(12\tau e^{\tau} + 42e^{\tau} - 30\tau^2 - 54\tau - 42\right).$$

Since $w_4(0) = 0$, we obtain

$$w_4(\tau) = 13\alpha^4 + \frac{19}{2}\alpha^4\tau + \frac{5}{2}\alpha^4\tau^2 + \frac{7}{2}\alpha^4\tau e^{\tau} + \frac{1}{2}\alpha^4\tau^2 e^{\tau} - 13\alpha^4 e^{\tau}.$$

The following recursive formula holds:

$$w_{2k}(\tau) = \frac{(-1)^k}{2^k k!}\alpha^{2k}\tau^{2k} + \int_0^{\tau} w_{2k}(s) ds - \int_0^{\tau} \frac{\alpha^2(\tau-s)^2}{2} w_{2k-2}(s)\, ds$$

$$+\int_0^{\tau} \frac{\alpha^4(\tau-s)^4}{2\cdot 2!} w_{2k-4}(s)\, ds - \int_0^{\tau} \frac{\alpha^6(\tau-s)^6}{2\cdot 3!} w_{2k-6}(s)\, ds + \cdots.$$

Taking into account equation [1.29], we have

$$h(t) = e^{-\frac{1}{2}\tau^2\alpha^2\varepsilon^2} + \varepsilon^2 \int_0^{\tau} e^{-\frac{1}{2}(\tau-s)^2\alpha^2\varepsilon^2} e^s ds$$

$$+\varepsilon^4 \int_0^{\tau} e^{-\frac{1}{2}(\tau-s)^2\alpha^2\varepsilon^2} \left(\frac{1}{2}\alpha^2 e^s - \alpha^2 - \alpha^2 s e^s\right) ds + \cdots.$$

Consider the following approximation:

$$h(t) = e^{-\frac{1}{2}\tau^2\alpha^2\varepsilon^2} + \varepsilon^2 \int_0^\tau e^{-\frac{1}{2}(\tau-s)^2\alpha^2\varepsilon^2} e^s ds.$$

After obtaining the inverse three-dimensional Fourier transform with respect to α (by using the Hankel transform), we get

$$g(\tau, x) = \mathcal{F}^{-1}(h(\tau)) = \frac{1}{2\pi^2\tau^3} e^{-\frac{x^2}{2\tau^2}} + \frac{\varepsilon^2}{2\pi^2} \int_0^\tau \frac{1}{(\tau-s)^3} e^{-\frac{x^2}{2(\tau-s)^2}} e^s ds$$

$$= \frac{1}{\sqrt{8\pi^3}\tau^3} e^{-\frac{x^2}{2\tau^2}} + \frac{\varepsilon^2}{\sqrt{8\pi^3}} e^\tau \int_0^\tau \frac{1}{u^3} e^{-\frac{x^2}{2u^2}} e^{-u} du + o(\varepsilon^2).$$

Then, for the pdf $f(t, x)$ of particle position at time t we have

$$f(t, x) = e^{-\lambda t} g(\lambda t, x) = \frac{e^{-\lambda t}}{\sqrt{8\pi^3}(\lambda t)^3} e^{-\frac{x^2}{2(\lambda t)^2}}$$

$$+ \frac{1}{\sqrt{8\pi^3}\lambda^2} \int_0^\infty \frac{1}{u^3} e^{-\frac{x^2}{2u^2}} e^{-u} du + o\left(\frac{1}{\lambda^2}\right).$$

$$\mathcal{F}^{-1}\left(e^{\frac{1}{2}t^2\alpha^2}\right) = \frac{1}{\sqrt{8\pi^3}} \int_0^\infty e^{-\frac{1}{2}t^2\alpha^2} \alpha^2 \frac{J_{\frac{1}{2}}(\alpha x)}{\sqrt{\alpha x}} d\alpha = \frac{1}{\sqrt{8\pi^3}} \frac{1}{t^3} e^{-\frac{x^2}{2t^2}}.$$

1.2.5. *Four-dimensional case*

We have that $\varphi(t) = \frac{J_1(\alpha t v)}{\alpha t v}$ when $v = const > 0$ in the four-dimensional case.

Now, let us assume that the velocity v_j is a random variable with the pdf

$$z(v) = \frac{v}{\sqrt{1-v^2}}, \quad 0 \le v < 1.$$

From [1.14], it follows that

$$\varphi(t) = \int_0^1 \frac{J_1(\alpha t v)}{\alpha t v} \frac{v}{\sqrt{1-v^2}} dv = 2\left(\frac{\sin(\alpha t/2)}{\alpha t}\right)^2,$$

with the corresponding renewal equation

$$H(t) = e^{-\lambda t}\left(\frac{\sin(\alpha t/2)}{\alpha t}\right)^2$$

$$+2\lambda \int_0^t e^{-\lambda(t-u)} \left(\frac{\sin\left(\frac{\alpha}{2}(t-u)\right)}{\alpha(t-u)}\right)^2 H(u) du, \qquad [1.34]$$

for the exponential pdf $g(t) = \lambda e^{-\lambda t}\mathbb{I}_{\{t \ge 0\}}$.

We cannot solve equation [1.34] but it is easily seen that

$$f(t,\mathbf{x}) = \mathcal{F}^{-1}(H(t)) \geq e^{-\lambda t}\mathcal{F}^{-1}\left(\left(\frac{\sin(\alpha t/2)}{\alpha t}\right)^2\right).$$

We have, by using the Hankel transform (Bochner and Chandrasekhar 1949), that

$$\mathcal{F}^{-1}\left(\left(\frac{\sin(\alpha t/2)}{\alpha t}\right)^2\right) = \frac{1}{4\pi^2}\int_0^\infty \frac{J_1(\alpha t x)}{\alpha t x}\alpha^3\left(\frac{\sin(\alpha t/2)}{\alpha t}\right)^2 d\alpha$$

$$= \frac{1}{4\pi^2 t x^2}\frac{1}{\sqrt{t^2-x^2}}, \quad x<t, \quad x=|\mathbf{x}|, \quad \mathbf{x}=(x_1,x_2,x_3,x_4).$$

Hence,

$$f(t,\mathbf{x}) \to \infty, \quad x \to t.$$

This explosive effect phenomenon was not observed in the four-dimensional case for constant velocity, for the absolute continuous part of the distribution of the particle position.

1.3. The distribution of random motion at non-constant velocity in semi-Markov media

This section deals with random motion at a non-constant speed, with uniformly distributed directions where the direction alternations occur according to renewal epochs having a general distribution. We derive the renewal equation for the characteristic function of the transition density of the multidimensional motion. By using the renewal equation, we study the behavior of the transition density near the sphere of its singularity, for two- and four-dimensional cases and variable velocity and the three-dimensional case for constant velocity. As an example, we have derived the distribution for one-, two- and three-dimensional random motion.

As in the case of one-dimensional random motion, many of the papers on random motion with uniformly distributed directions on the multidimensional space are devoted to the analysis of models where the motions are driven by a homogeneous Poisson process, so their processes are Markov. In various studies (Di Crescenzo 2001; Pogorui 2011), a non-Markov generalization of one-dimensional random evolutions of the telegrapher's random process, where the motion is driven by an alternating semi-Markov process with Erlang-distributed interrenewal times, is considered. Random flights in $\mathbb{R}^n$, with K-Erlang-distributed displacements and uniformly distributed directions, have been studied in Pogorui and Rodríguez-Dagnino (2011). A planar random motion performed by a particle that changes direction at even-valued Poisson events is studied in Beghin and Orsingher (2010a). In previous studies (Le Caër 2010, 2011), a random walk with steps of a uniform orientation and Dirichlet-distributed lengths is considered. The transition density, which has simple analytical forms for two- and four-dimensional Markov random motion, has been derived by Orsingher and De Gregorio (2007). In this work, we consider multidimensional random motions with uniformly distributed directions, with general distributed steps and non-constant velocity, and we have extended some results of Pogorui and Rodríguez-Dagnino (2005), Pogorui (2011) and Pogorui and Rodríguez-Dagnino (2011). We show some interesting solvable cases that have not been reported by other studies.

Let us consider the renewal process $\nu(t) = \max\{m \geq 0 : \ \tau_m \leq t\}, t \geq 0$, where $\tau_m = \sum_{k=1}^{m} \theta_k$, $\tau_0 = 0$ and $\theta_k \geq 0$, $k = 1, 2, \ldots$, are i.i.d. with a distribution function $G(t)$ and the probability density function $g(t) = \frac{d}{dt}G(t)$.

We assume that a particle starting from the coordinate origin $(0, 0, \ldots, 0)$ of the space R^n, at time $t = 0$, continues its motion with a velocity $v(s) > 0$, $0 \leq s < \theta_1$ along the direction of $\boldsymbol{\eta}_0^{(n)}$, where $\boldsymbol{\eta}_0^{(n)} = (x_1, x_2, \cdots, x_n)$ is a random n-dimensional vector uniformly distributed on the unit sphere

$$\Omega_1^{n-1} = \left\{(x_1, x_2, \cdots, x_n) : \ x_1^2 + x_2^2 + \cdots + x_n^2 = 1\right\}.$$

At instant τ_1, the particle changes its direction to $\boldsymbol{\eta}_1^{(n)}$, where $\boldsymbol{\eta}_1^{(n)}$ and $\boldsymbol{\eta}_0^{(n)}$ are independent and identically distributed on Ω_1^{n-1}, and continues its motion with the velocity $v(s)$, $0 \leq s < \theta_2$ along the direction of $\boldsymbol{\eta}_1^{(n)}$. Then at instant τ_2, the particle changes its direction to $\boldsymbol{\eta}_2^{(n)}$, where $\boldsymbol{\eta}_2^{(n)}$ is also uniformly distributed on Ω_1^{n-1} and independent of $\boldsymbol{\eta}_0^{(n)}$, $\boldsymbol{\eta}_1^{(n)}$, and continues its motion with the velocity $v(s)$, $0 \leq s < \theta_3$ along the direction of $\boldsymbol{\eta}_2^{(n)}$, and so on.

Denote by $\mathbf{x}^{(n)}(t)$, $t \geq 0$ the particle position at time t. We have that

$$\mathbf{x}^{(n)}(t) = \sum_{j=1}^{\nu(t)} \boldsymbol{\eta}_{j-1}^{(n)} \int_0^{\theta_j} v(s)\, ds + \boldsymbol{\eta}_{\nu(t)}^{(n)} \int_0^{t-\tau_{\nu(t)}} v(s)\, ds, \qquad [1.35]$$

where we assume that $\sum_{j=1}^{0} \boldsymbol{\eta}_{j-1}^{(n)} \int_0^{\theta_j} v(s)\, ds = 0$.

Basically, equation [1.35] determines the random evolution in the semi-Markov medium $\nu(t)$.

The counting process $\{\nu(t)\}$ gives the number of velocity alternations that have occurred in the interval $(0, t)$.

The probabilistic properties of a random vector $\mathbf{x}^{(n)}(t)$ are completely determined by its projection $x^{(n)}(t) = \sum_{j=1}^{\nu(t)} \eta_{j-1}^{(n)} \int_0^{\theta_j} v(s)\, ds + \eta_{\nu(t)}^{(n)} \int_0^{t-\tau_{\nu(t)}} v(s)\, ds$ on a fixed line, where $\eta_j^{(n)}$ is the projection of $\boldsymbol{\eta}_j^{(n)}$ on the line.

Indeed, let us consider the distribution function $F_x(y) = P\left(x^{(n)}(t) \leq y\right)$. Then, the characteristic function $H(t)$ of $\mathbf{x}^{(n)}(t)$ is given by:

$$\begin{aligned} H(t) &= E\exp\left\{i\left(\boldsymbol{\alpha}, \mathbf{x}^{(n)}(t)\right)\right\} = E\exp\left\{i\left\|\boldsymbol{\alpha}\right\|\left(\mathbf{e}, \mathbf{x}^{(n)}(t)\right)\right\} \\ &= E\exp\left\{i\left\|\boldsymbol{\alpha}\right\| x^{(n)}(t)\right\} = \int_0^{\infty} \exp\left\{i\left\|\boldsymbol{\alpha}\right\| y\right\} dF_x(y), \end{aligned}$$

where $\|\boldsymbol{\alpha}\| = \sqrt{\alpha_1^2 + \alpha_2^2 + \cdots + \alpha_n^2}$.

Let us denote by $f_{\eta^{(n)}}(x)$, the pdf of the projection $\eta_j^{(n)}$ of vector $\boldsymbol{\eta}_j^{(n)}$ onto a fixed line. In Pogorui (2010b), we proved that:

$$f_{\eta^{(n)}}(x) = \begin{cases} \frac{\Gamma\left(\frac{n}{2}\right)}{\sqrt{\pi}\Gamma\left(\frac{n-1}{2}\right)}\left(1-x^2\right)^{(n-3)/2}, & x \in [-1,1], \\ 0, & x \notin [-1,1], \end{cases} \qquad [1.36]$$

and

$$\begin{aligned} \varphi(t,\alpha) &= \varphi_{\eta^{(n)}}\left(\alpha \int_0^t v(s)\, ds\right) = \mathrm{E}\left[e^{i\int_0^t v(s)ds\left(\boldsymbol{\alpha}, \boldsymbol{\eta}_j^{(n)}\right)}\right] \\ &= \int_{-1}^{1} e^{i\alpha\int_0^t v(s)dsx} f_{\eta^{(n)}}(x)dx, \end{aligned}$$

where $\alpha = \|\boldsymbol{\alpha}\|$.

It is not hard to verify that

$$\varphi(t,\alpha) = 2^{\frac{n-2}{2}}\Gamma\left(\frac{n}{2}\right)\frac{J_{\frac{n-2}{2}}\left(\alpha\int_0^t v(s)\, ds\right)}{\left(\alpha\int_0^t v(s)\, ds\right)^{\frac{n-2}{2}}}.$$

1.3.1. *Renewal equation for the characteristic function*

THEOREM 1.4.– The characteristic function $H(t) = \mathbf{E}\left[\exp\left\{i\left(\boldsymbol{\alpha}, \mathbf{x}^{(n)}(t)\right)\right\}\right], t \geq 0$, of the random motion $\mathbf{x}^{(n)}(t)$ is a solution of the following Volterra integral equation:

$$H(t) = (1 - G(t))\,\varphi(t,\alpha) + \int_0^t g(u)\,\varphi(u,\alpha)\, H(t-u)\, du. \qquad [1.37]$$

PROOF.– It follows from equation [1.35] that

$$H(t) = \mathbf{E}\left[\exp\left\{i\left(\boldsymbol{\alpha}, \mathbf{x}^{(n)}(t)\right)\right\}\right]$$

$$= \mathbf{E}\left[\exp\left\{i\left(\boldsymbol{\alpha}, \sum_{j=1}^{\nu(t)} \boldsymbol{\eta}_{j-1}^{(n)} \int_0^{\theta_j} v(s)\, ds + \boldsymbol{\eta}_{\nu(t)}^{(n)} \int_0^{t-\tau_{\nu(t)}} v(s)\, ds\right)\right\}\right]$$

$$= (1 - G(t))\, \mathbf{E}\left[e^{i \int_0^t v(s) ds \left(\boldsymbol{\alpha}, \boldsymbol{\eta}_0^{(n)}\right)}\right]$$

$$+ \int_0^t g(u)\, \mathbf{E}\left[e^{i \int_0^u v(s) ds \left(\boldsymbol{\alpha}, \boldsymbol{\eta}_0^{(n)}\right)}\right] H(t-u)\, du.$$

To complete the proof, observe that $\varphi(t, \alpha) = \mathbf{E}\left[e^{i \int_0^t v(s) ds \left(\boldsymbol{\alpha}, \boldsymbol{\eta}_0^{(n)}\right)}\right]$.

Denote by $f_n(t, \mathbf{x})$, the pdf of the particle position at time t. It is easily seen that $f_n(t, \mathbf{x}) = \mathcal{F}^{-1}(H(t))$.

Our purpose is to study $f_n(t, \mathbf{x})$. It is not hard to verify that $f_n(t, \mathbf{x})$ is a radial function, i.e. $f_n(t, \mathbf{x}) = f_n(t, \mathbf{y})$, for $\|\mathbf{x}\| = \|\mathbf{y}\|$ and we will also use the notation $f_n(t, x)$, where $x = \|\mathbf{x}\|$.

1.3.2. *Two-dimensional case*

In the two-dimensional case, $\varphi(t, \alpha) = J_0\left(\alpha \int_0^t v(s)\, ds\right)$. Let us consider the case where $g(t) = \lambda e^{-\lambda t} \mathbb{I}_{\{t \geq 0\}}$, $v(s) = \frac{c}{2\sqrt{s}}$, $c > 0$. Therefore, $\varphi(t, \alpha) = J_0\left(\alpha c\sqrt{t}\right)$ and we have

$$H(t) = e^{-\lambda t} J_0\left(\alpha c\sqrt{t}\right) + \int_0^t \lambda e^{-\lambda u} J_0\left(\alpha c\sqrt{u}\right) H(t-u)\, du.$$

We first define $h(t) = e^{\lambda t} H(t)$, and then we obtain

$$h(t) = J_0\left(\alpha c\sqrt{t}\right) + \lambda \int_0^t J_0\left(\alpha c\sqrt{u}\right) h(t-u)\, du. \qquad [1.38]$$

Denote by $\hat{h}(p) = \mathcal{L}(h(t), p) = \int_0^\infty e^{-tp} h(t) dt$, the Laplace transform of $h(t)$.

By applying the Laplace transform in equation [1.38], we have

$$\hat{h}(p) = \frac{1}{p} e^{-\frac{\alpha^2 c^2}{4p}} + \frac{\lambda}{p} e^{-\frac{\alpha^2 c^2}{4p}} \hat{h}(p).$$

Hence,

$$\hat{h}(p) = \frac{1}{\lambda}\sum_{n=1}^{\infty}\left(\frac{\lambda}{p}e^{-\frac{\alpha^2c^2}{4p}}\right)^n.$$

Since this series converges uniformly for $Re(p) \geq \lambda + \sigma$, where σ is a fixed positive number, we can obtain the inverse Laplace, which is given by

$$h(t) = \sum_{n=1}^{\infty}\left(\frac{4t\lambda^2}{n\alpha^2c^2}\right)^{\frac{n-1}{2}} J_{n-1}(\alpha c\sqrt{tn})$$

or equivalently

$$H(t) = e^{-\lambda t}\sum_{n=1}^{\infty}\left(\frac{4t\lambda^2}{n\alpha^2c^2}\right)^{\frac{n-1}{2}} J_{n-1}(\alpha c\sqrt{tn}).$$

It is well known that

$$\mathcal{F}^{-1}\left(J_0\left(\alpha c\sqrt{t}\right)\right) = \frac{1}{2\pi}\delta\left(c^2t - x^2\right),$$

where $\mathcal{F}^{-1}$ is the two-dimensional inverse Fourier transform, and $\delta(\cdot)$ is the Dirac delta function.

To calculate the two-dimensional inverse Fourier transform of $H(t)$, we use the Hankel transform as follows:

$$\frac{1}{2\pi}\int_0^{\infty}\left(\frac{4t\lambda^2}{n\alpha^2c^2}\right)^{\frac{n-1}{2}} \alpha J_{n-1}\left(\alpha c\sqrt{tn}\right) J_0(\alpha x)\, d\alpha$$

$$= \frac{1}{\pi}\frac{\lambda^{n-1}\left(nc^2t - x^2\right)^{n-2}}{\Gamma(n-1)\, n^{n-1}c^{2n-1}}, \quad n \geq 2.$$

It is well known (Di Crescenzo 2001; Orsingher and De Gregorio 2007) that

$$f_2(t, \mathbf{x}) = s_2(t, \mathbf{x}) + c_2(t, \mathbf{x}),$$

where $s_2(t, \mathbf{x})$ is the singular part and $c_2(t, \mathbf{x})$ is the continuous part of the pdf $f_2(t, \mathbf{x})$.

It is not hard to verify that

$$s_2(t, \mathbf{x}) = \frac{1}{2\pi}\delta\left(c^2t - x^2\right)$$

and

$$c_2(t,\mathbf{x}) = \frac{e^{-\lambda t}}{\pi}\sum_{n=2}^{\infty}\frac{\lambda^{n-1}\left(nc^2t - x^2\right)^{n-2}}{\Gamma(n-1)\,n^{n-1}c^{2n-1}}.$$

Hence, for $c^2t \geq x^2$, we have

$$f_2(t,\mathbf{x}) = \mathcal{F}^{-1}(H(t),\alpha) = \frac{e^{-\lambda t}}{2\pi}\delta\left(c^2t - x^2\right) + \frac{e^{-\lambda t}}{\pi}\sum_{n=2}^{\infty}\frac{\lambda^{n-1}\left(nc^2t - x^2\right)^{n-2}}{\Gamma(n-1)\,n^{n-1}c^{2n-1}}.$$

It is easily seen that the series $c_2(t,\mathbf{x})$ converges uniformly for $|\mathbf{x}| \leq c\sqrt{t}$ and there is no explosion effect as $|\mathbf{x}| \to c\sqrt{t}$, i.e.

$$\begin{aligned}\lim_{|\mathbf{x}|\to c\sqrt{t}} c_2(t,\mathbf{x}) &= \lim_{|\mathbf{x}|\to c\sqrt{t}}\frac{e^{-\lambda t}}{\pi}\sum_{n=2}^{\infty}\frac{\lambda^{n-1}\left(nc^2t - x^2\right)^{n-2}}{\Gamma(n-1)\,n^{n-1}c^{2n-1}}\\ &= \frac{e^{-\lambda t}}{\pi}\sum_{n=2}^{\infty}\frac{\lambda^{n-1}\left((n-1)\,c^2t\right)^{n-2}}{\Gamma(n-1)\,n^{n-1}c^{2n-1}}\\ &= \frac{e^{-\lambda t}}{\pi c^3}\sum_{n=2}^{\infty}\frac{\lambda^{n-1}\left(\left(1-\frac{1}{n}\right)t\right)^{n-2}}{\Gamma(n-1)\,n} < +\infty.\end{aligned}$$

Therefore, $c_2(t,\mathbf{x})$ does not go to infinity, as $|\mathbf{x}|$ goes to the circle of singularity $C\left(0,c\sqrt{t}\right) = \left\{\mathbf{x} = (x_1,x_2) \middle| \, x_1^2 + x_2^2 = c^2t\right\}$ of $f_2(t,\mathbf{x})$, which takes place for the movement of a particle with a constant velocity on a plane (Orsingher and De Gregorio 2007). The absence of the explosion effect can be explained by the consideration that the particle velocity $v(s) = \frac{c}{2\sqrt{s}}$, which is very large just after renewal epoch, makes the particle leaves very quickly from the layer closest to the singularity area.

1.3.3. *Three-dimensional case*

It was proved in Kolesnik and Pinsky (2011), that in the case of $g(t) = \lambda e^{-\lambda t}\mathbb{I}\{t \geq 0\}$ and $v(s) = v = constant$, the n-dimensional random evolutions ($n \geq 3$) are driven by the hyperparabolic operators composed of the telegraph operators and their integer powers. However, it is not easy to find closed-form solutions in this case.

We consider the case where $g(t)$ is Erlang-2 pdf and $v(s) = v = const$. We will show that these three-dimensional random evolutions are driven by the three-dimensional telegraph equation.

Let us consider the case where $g(t) = \lambda^2 te^{-\lambda t}\mathbb{I}\{t \geq 0\}$ and $v(s) = v = const.$ Thus, we have

$$\varphi(t,\alpha) = \frac{\sin(\alpha vt)}{\alpha vt}, \qquad \alpha = \sqrt{\alpha_1^2 + \alpha_2^2 + \alpha_3^2}$$

Therefore,

$$H(t) = \left(e^{-\lambda t} + \lambda te^{-\lambda t}\right)\frac{\sin(\alpha vt)}{\alpha vt}$$
$$+\int_0^t \lambda^2 (t-u) e^{-\lambda(t-u)}\frac{\sin(\alpha v(t-u))}{\alpha v(t-u)} H(u)\,du$$
$$= \left(e^{-\lambda t} + \lambda te^{-\lambda t}\right)\frac{\sin(\alpha vt)}{\alpha vt}$$
$$+\frac{\lambda^2}{\alpha v}\int_0^t e^{-\lambda(t-u)}\sin(\alpha v(t-u))H(u)\,du. \qquad [1.39]$$

The pdf $f_3(t,\mathbf{x}) = \mathcal{F}^{-1}(H(t),\alpha)$, where $\mathcal{F}^{-1}$ is the three-dimensional inverse Fourier transform with respect to α, is a radial function (i.e. it depends only on $x = \sqrt{x_1^2 + x_2^2 + x_3^2}$), and it has the following form:

$$f_3(t,x) = s_3(t,x) + c_3(t,x),$$

where $s_3(t,x)$ is the singular part and $c_3(t,x)$ is the continuous part of the pdf $f_3(t,x)$.

It is well known that the function $4\pi R\frac{\sin \alpha R}{\alpha}$ is the Fourier transform of the simple layer $\delta_{S(R)}$, (Vladimirov 1996, p. 154). Hence,

$$s_3(t,x) = \mathcal{F}^{-1}\left(\left(e^{-\lambda t} + \lambda te^{-\lambda t}\right)\frac{\sin(\alpha vt)}{\alpha vt}\right) = \left(e^{-\lambda t} + \lambda te^{-\lambda t}\right)\frac{\delta_{S(vt)}}{4\pi (vt)^2}.$$

Denote by $H_s(t) = \int_{\mathbb{R}^3} e^{i(\boldsymbol{\alpha},\mathbf{x})}s_3(t,\mathbf{x})d\mathbf{x}$ and by $H_c(t) = \int_{\mathbb{R}^3} e^{i(\boldsymbol{\alpha},\mathbf{x})}c_3(t,\mathbf{x})d\mathbf{x}$. It is easily seen that $H(t) = H_s(t) + H_c(t)$.

Since $H_s(t) = \left(e^{-\lambda t} + \lambda te^{-\lambda t}\right)\frac{\sin(\alpha vt)}{\alpha vt}$ from equation [1.39], then it follows that

$$H_c(t) = \frac{\lambda^2}{\alpha v}\int_0^t e^{-\lambda(t-u)}\sin(\alpha v(t-u))\;(H_s(u) + H_c(u))\,du$$

or equivalently,

$$H_c(t) = \frac{\lambda^2}{\alpha v} e^{-\lambda t} \int_0^t \sin(\alpha v (t-u)) \, (1+\lambda u) \frac{\sin(\alpha v u)}{\alpha v u} du$$
$$+ \frac{\lambda^2}{\alpha v} \int_0^t e^{-\lambda(t-u)} \sin(\alpha v (t-u)) \, H_c(u) \, du.$$

Let us consider

$$h(t) = \lambda^2 \int_0^t \frac{\sin(\alpha v (t-u))}{\alpha v} (1+\lambda u) \frac{\sin(\alpha v u)}{\alpha v u} du$$
$$+ \frac{\lambda^2}{\alpha v} \int_0^t \sin(\alpha v (t-u)) \, h(u) \, du, \qquad [1.40]$$

where $h(t) = e^{\lambda t} H_c(t)$.

Differentiating equation [1.40] with respect to t two times, we obtain

$$\frac{dh(t)}{dt} - \lambda^2 \int_0^t \cos(\alpha v (t-u)) \, h(u) \, du$$
$$= \lambda^2 \int_0^t \cos(\alpha v (t-u)) \, (1+\lambda u) \frac{\sin(\alpha v u)}{\alpha v u} du,$$

and

$$\frac{d^2 h(t)}{dt} - \lambda^2 h(t) + \lambda^2 \alpha v \int_0^t \sin(\alpha v (t-u)) \, h(u) \, du$$
$$= (\lambda^2 + \lambda^3 t) \frac{\sin(\alpha v t)}{\alpha v t} - \lambda^2 \int_0^t \sin(\alpha v (t-u)) \, (1+\lambda u) \frac{\sin(\alpha v u)}{u} du. \qquad [1.41]$$

Combining the equations [1.40] and [1.41], we obtain

$$\frac{d^2 h(t)}{dt} - \lambda^2 h(t) + \alpha^2 v^2 h(t) = (\lambda^2 + \lambda^3 t) \frac{\sin(\alpha v t)}{\alpha v t}. \qquad [1.42]$$

Denote by $g(t, \mathbf{x}) = \mathcal{F}^{-1}(h(t), \alpha)$ (recall that $h(t)$ depends on α), the three-dimensional inverse Fourier transform. Here, we use the notation $g(t, \mathbf{x})$, and we should remember that g is a radial function and $g(t, \mathbf{x}) = g(t, \mathbf{y})$ for all $\mathbf{x}$, $\mathbf{y} \in \mathbb{R}^3$, such that $|\mathbf{x}| = |\mathbf{y}|$.

After obtaining the three-dimensional inverse Fourier transform of equation [1.42], for $0 \le x \le vt$ we have

$$\frac{\partial^2 g(t, \mathbf{x})}{\partial t^2} - \lambda^2 g(t, \mathbf{x}) - v^2 \sum_{i,j=1}^{3} \frac{\partial^2 g(t, \mathbf{x})}{\partial x_i \partial x_j} = (\lambda^2 + \lambda^3 t) \frac{\delta_{S(vt)}}{4\pi (vt)^2}. \qquad [1.43]$$

Since $e^{\lambda t} c_3(t, \mathbf{x}) = \mathcal{F}^{-1}(H_c(t), \alpha) = g(t, \mathbf{x})$ from equation [1.43], for $0 \leq x \leq vt$ we obtain

$$\frac{\partial^2 c_3(t,\mathbf{x})}{\partial t^2} + 2\lambda \frac{\partial}{\partial t} c_3(t,\mathbf{x}) - v^2 \sum_{i,j=1}^{3} \frac{\partial^2 c_3(t,\mathbf{x})}{\partial x_i \partial x_j}$$

$$= \left(\lambda^2 + \lambda^3 t\right) e^{-\lambda t} \frac{\delta_{S(vt)}}{4\pi (vt)^2}, \qquad [1.44]$$

with the initial conditions $c_3(t, \mathbf{x}) = 0$ and $\left.\frac{\partial c_3(t,\mathbf{x})}{\partial t}\right|_{t=0} = 0$, which can be obtained directly from equation [1.40]. We have $s_3(t, \mathbf{x}) = \left(e^{-\lambda t} + \lambda t e^{-\lambda t}\right) \frac{\delta_{S(vt)}}{4\pi(vt)^2}$ by solving equation [1.44]. Therefore, we obtain the pdf $f_3(t, \mathbf{x})$.

REMARK 1.2.– *It was the subject of discussion among researchers on whether the multidimensional random flights could be described by the telegraph equations, similarly to the one-dimensional case (Kolesnik and Pinsky 2011). We think that for the three-dimensional case, equation [1.44] answers this question. Then, it seems natural to call equation [1.44] the generalized Goldstein–Kac equation in three dimensions.*

1.3.4. *Four-dimensional case*

In this case

$$\varphi(t, \alpha) = \frac{J_1\left(\alpha \int_0^t v(s)\, ds\right)}{\alpha \int_0^t v(s)\, ds},$$

where J_1 is the Bessel function of order 1.

Let us consider the case for a 2-Erlang pdf, i.e. $g(t) = \lambda^2 t e^{-\lambda t} \mathbb{I}\{t \geq 0\}$ and velocity $v(s) = \frac{1}{2\sqrt{s}}$.

Therefore,

$$H(t) = \left(e^{-\lambda t} + \lambda t e^{-\lambda t}\right) \frac{J_1(\alpha\sqrt{t})}{\alpha\sqrt{t}} + \int_0^t \lambda^2 u e^{-\lambda u} \frac{J_1(\alpha\sqrt{u})}{\alpha\sqrt{u}} H(t-u)\, du,$$

or

$$h(t) = (1 + \lambda t) \frac{J_1(\alpha\sqrt{t})}{\alpha\sqrt{t}} + \frac{\lambda^2}{\alpha} \int_0^t J_1(\alpha\sqrt{u}) \sqrt{u}\, h(t-u)\, du, \qquad [1.45]$$

where $h(t) = H(t) e^{\lambda t}$.

After obtaining the Laplace transform in equation [1.45], we have

$$\hat{h}(p) = \frac{2 - 2e^{-\frac{\alpha^2}{4p}}}{\alpha^2} + \frac{\lambda e^{-\frac{\alpha^2}{4p}}}{2p^2} + \frac{\lambda^2 e^{-\frac{\alpha^2}{4p}}}{2p^2} \hat{h}(p).$$

Hence,

$$\hat{h}(p) = \frac{2 - 2e^{-\frac{\alpha^2}{4p}}}{\alpha^2} + \frac{\lambda e^{-\frac{\alpha^2}{4p}}}{2p^2} + \frac{2 - 2e^{-\frac{\alpha^2}{4p}}}{\alpha^2} \sum_{n=1}^{\infty} \left(\frac{\lambda^2 e^{-\frac{\alpha^2}{4p}}}{2p^2} \right)^n + \frac{1}{\lambda} \sum_{n=2}^{\infty} \left(\frac{\lambda^2 e^{-\frac{\alpha^2}{4p}}}{2p^2} \right)^n. \qquad [1.46]$$

By obtaining the Laplace transform in equation [1.46], we have

$$h(t) = (1 + \lambda t) \frac{J_1(\alpha\sqrt{t})}{\alpha\sqrt{t}} + \sum_{n=1}^{\infty} \lambda^{2n} t^{\frac{2n-1}{2}} 2^n \times \left(\frac{(n+1)^{n-\frac{1}{2}}}{n^{n-\frac{1}{2}}} \frac{J_{2n-1}(\sqrt{n}\alpha\sqrt{t})}{\alpha^{2n+1}} - \frac{n^{n-\frac{1}{2}}}{(n+1)^{n-\frac{1}{2}}} \frac{J_{2n-1}(\sqrt{n+1}\alpha\sqrt{t})}{\alpha^{2n+1}} \right) + \sum_{n=2}^{\infty} \frac{2^{n-1}\lambda^{2n-1}\sqrt{n}\, t^{\frac{2n-1}{2}}}{n^n} \frac{J_{2n-1}(\sqrt{n}\alpha\sqrt{t})}{\alpha^{2n-1}}. \qquad [1.47]$$

We will use the following Hankel transforms to calculate the four-dimensional inverse Fourier transforms (Vladimirov 1996, p. 321):

$$\int_0^{\infty} \frac{J_{2n-1}(\sqrt{n}\alpha\sqrt{t})}{\alpha^{2n+1}} \alpha^2 J_1(\alpha x)\, d\alpha = \frac{2(nt)^{n-1/2}\left(1 - \left(\frac{nt-x^2}{nt}\right)^{2n-1}\right)}{2^{2n} x \Gamma(2n)},$$

$$\int_0^{\infty} \frac{J_{2n-1}(\sqrt{n+1}\alpha\sqrt{t})}{\alpha^{2n+1}} \alpha^2 J_1(\alpha x) d\alpha = \frac{2(nt+t)^{n-1/2}\left(1 - \left(\frac{(n+1)t-x^2}{(n+1)t}\right)^{2n-1}\right)}{2^{2n} x \Gamma(2n)},$$

$$\int_0^{\infty} \frac{J_{2n-1}(\sqrt{n}\alpha\sqrt{t})}{\alpha^{2n-1}} \alpha^2 J_1(\alpha x)\, d\alpha = \frac{8x(nt - x^2)^{2n-3}}{4^n (tn)^{n-1/2} \Gamma(2n-2)}. \qquad [1.48]$$

Then, by considering equation [1.48] we can obtain the four-dimensional inverse Fourier transform in equation [1.47]. For instance, for $t \geq x^2$ we have

$$f_4(t, \mathbf{x}) = \mathcal{F}^{-1}(H(t, \alpha)) = \frac{\left(e^{-\lambda t} + \lambda t e^{-\lambda t}\right)}{2\pi} \delta\left(t - x^2\right)$$

$$+e^{-\lambda t} \sum_{n=1}^{\infty} \lambda^{2n} t^{\frac{2n-1}{2}} \left(\frac{2(nt+t)^{n-1/2}\left(1 - \left(\frac{nt-x^2}{nt}\right)^{2n-1}\right)}{2^n x \Gamma(2n)} \right.$$

$$\left. - \frac{2(nt)^{n-1/2}\left(1 - \left(\frac{(n+1)t-x^2}{(n+1)t}\right)^{2n-1}\right)}{2^n x\, \Gamma(2n)} \right)$$

$$+e^{-\lambda t} \sum_{n=2}^{\infty} \frac{\lambda^{2n-1} t^{\frac{2n-1}{2}} \sqrt{n}}{n^n} \frac{4x\left(nt - x^2\right)^{2n-3}}{2^n (tn)^{n-1/2} \Gamma(2n-2)}.$$

Therefore, the singular part of the pdf $f_4(t, \mathbf{x}) = s_4(t, \mathbf{x}) + c_4(t, \mathbf{x})$, is given by

$$s_4(t, \mathbf{x}) = \frac{\left(e^{-\lambda t} + \lambda t e^{-\lambda t}\right)}{2\pi} \delta\left(t - x^2\right)$$

and the continuous part of $f_4(t, \mathbf{x})$ is

$$c_4(t, \mathbf{x}) = e^{-\lambda t} \sum_{n=1}^{\infty} \lambda^{2n} t^{\frac{2n-1}{2}}$$

$$\times \left(\frac{2(nt+t)^{n-1/2}\left(1 - \left(\frac{nt-x^2}{nt}\right)^{2n-1}\right)}{2^n x \Gamma(2n)} - \frac{2(nt)^{n-1/2}\left(1 - \left(\frac{(n+1)t-x^2}{(n+1)t}\right)^{2n-1}\right)}{2^n x \Gamma(2n)} \right)$$

$$+e^{-\lambda t} \sum_{n=2}^{\infty} \frac{\lambda^{2n-1} t^{\frac{2n-1}{2}} \sqrt{n}}{n^n} \frac{4x\left(nt - x^2\right)^{2n-3}}{2^n (tn)^{n-1/2} \Gamma(2n-2)}.$$

It is easily verified that the series $c_4(t, \mathbf{x})$ converges uniformly for $|\mathbf{x}| \leq \sqrt{t}$ and that there is no explosion effect, as $|\mathbf{x}| \to \sqrt{t}$, i.e.

$$\lim_{|\mathbf{x}| \to \sqrt{t}} c_4(t, \mathbf{x}) = \lim_{|z| \to \sqrt{t}} e^{-\lambda t} \sum_{n=1}^{\infty} \lambda^{2n} t^{\frac{2n-1}{2}}$$

$$\times \left(\frac{2(nt+t)^{n-1/2}\left(1 - \left(\frac{nt-x^2}{nt}\right)^{2n-1}\right)}{2^n x \Gamma(2n)} \right.$$

$$-\frac{2(nt)^{n-1/2}\left(1-\left(\frac{(n+1)t-x^2}{(n+1)t}\right)^{2n-1}\right)}{2^n x\,\Gamma(2n)}\Bigg)$$

$$+e^{-\lambda t}\sum_{n=2}^{\infty}\frac{\lambda^{2n-1}t^{\frac{2n-1}{2}}\sqrt{n}}{n^n}\frac{4x\left(nt-x^2\right)^{2n-3}}{2^n(tn)^{n-1/2}\Gamma(2n-2)}$$

$$=e^{-\lambda t}\sum_{n=1}^{\infty}\lambda^{2n}t^{n-1}\left(\frac{2(nt+t)^{n-1/2}\left(1-\left(\frac{nt-t}{nt}\right)^{2n-1}\right)}{2^n\Gamma(2n)}\right.$$

$$\left.-\frac{2(nt)^{n-1}\left(1-\left(\frac{nt}{(n+1)t}\right)^{2n-1}\right)}{2^n\sqrt{n}\,\Gamma(2n)}\right)$$

$$+e^{-\lambda t}\sum_{n=2}^{\infty}\frac{\lambda^{2n-1}t^n\sqrt{n}}{n^n}\frac{4(nt-t)^{2n-3}}{2^n(tn)^{n-1/2}\Gamma(2n-2)}<+\infty.$$

Therefore, $c_4(t,\mathbf{x})$ does not go to infinity as $|x|$ goes to the circus of singularity $C\left(0,c\sqrt{t}\right)=\left\{\mathbf{x}=(x_1,x_2,x_3,x_4)|\,x_1^2+x_2^2+x_3^2+x_4^2=c^2t\right\}$ of $f_4(t,\mathbf{x})$.

1.4. Goldstein–Kac telegraph equations and random flights in higher dimensions

In this section, we deal with random motions in dimensions two, three, and five, where the governing equations are telegraph-type equations. Our methodology is first applied to the second-order telegraph equation, then we refine the well-known results found by other methods. Next, we show that the (2,λ)-Erlang distribution for sojourn times defines the underlying stochastic process for the three-dimensional Goldstein–Kac type telegraph equation. By finding the corresponding fundamental solution of this equation, we have obtained the approximated expression for the transition density of the three-dimensional movement. Our results are more complete than previous ones, and may have important consequences in applications. We also obtain the five-dimensional telegraph-type equation by assuming a random motion with a (4,λ)-Erlang distribution for sojourn times. Such an equation can be factorized as the product of two telegraph-type equations, where one of them is the Goldstein–Kac five-dimensional telegraph equation. In our analysis, the dimension n is related to the ($n-1$,λ)-Erlang distribution for sojourn times of the random walks.

We should remember that the problem of a one-dimensional random motion, defined by the movement of a particle at a constant velocity v, traveling in a direction for a random distance drawn from an exponential distribution, and the particle change to the opposite direction under the same stochastic conditions, was studied and solved by Goldstein (1951) and Kac (1974) in 1951 and 1956, respectively. This problem is specified as a random motion governed by a switching Poisson process with alternating directions, but having exponentially distributed sojourn times, i.e. it is essentially a Markov process. The solution of this problem satisfies the one-dimensional telegraph-type equation (this equation has a similar form as the Heaviside telegraph equation for wave propagation in transmission lines)

$$\frac{\partial^2 f(t,x)}{\partial t^2} + 2\lambda\frac{\partial}{\partial t}f(t,x) = v^2\frac{\partial^2 f(t,x)}{\partial x^2}, \qquad [1.49]$$

and it is called the Goldstein–Kac telegraph equation. Kac (1974) states the possibility of extending the same analysis for modeling multidimensional random motions, satisfying, Goldstein–Kac telegraph equations in higher dimensions. In the last paragraph of his paper, Kac states:

> This same proof goes also for a higher number of dimensions. Again it is simply a matter of writing the Laplace transform and verifying the same formula.

We should mention that the problem of assessing the validity of this statement has been an elusive problem for researchers working in this area since then, i.e. to find underlying stochastic processes describing multidimensional random evolutions and that satisfy telegraph-type equations. Hitherto there is a definitive answer to this problem, so we cannot ensure that multidimensional random flights can be described by multidimensional telegraph equations, similarly to the one-dimensional case, i.e. equations of the following type:

$$\frac{\partial^2 f(t,\mathbf{x})}{\partial t^2} + 2\lambda\frac{\partial}{\partial t}f(t,\mathbf{x}) = v^2\sum_{i=1}^{n}\frac{\partial^2 f(t,\mathbf{x})}{\partial x_i^2}, \qquad [1.50]$$

where $\mathbf{x} = (x_1, x_2, \ldots, x_n) \in \mathbb{R}^n$, $n \geq 2$.

Some remarkable efforts in this direction are presented in De Gregorio and Orsingher (2012) and Garra and Orsingher (2014), where they show that for random motions in three dimensions, with particles that only change direction at even-valued Poisson events, the governing partial differential equation is a Goldstein–Kac telegraph equation. However, the solution presented by these authors seems to be in complete and it is not a true pdf.

In this work, we provide the governing partial differential equations for random evolutions in two, three and five dimensions. We focus on these two odd dimension cases since the first one is very important in applications. We are also including the two-dimensional random flight, whose transition, density is the fundamental solution to the two-dimensional Goldstein-Kac telegraph equation. This problem has been studied by several researchers in the past and our main purpose is to show that it is also consistent with our assumption of having n-dimensional random flights with $(n-1, \lambda)$-Erlang-distributed sojourn times. Furthermore, we are adding some new insights to the solution of this two-dimensional random flight.

1.4.1. *Preliminaries about our modeling approach*

Let $\nu(t)$ be the renewal process that starts with $\nu(0) = 0$. We define the time of the kth jump by $\tau_k = \inf\{t : \nu(t) = k\}$, $k \geq 0$. The random variables τ_k are called jump times, $\theta_k = \tau_{k+1} - \tau_k$ is called the kth renewal interval (or interarrival time) and θ_k, $k \in$ are non-negative, independent, identically distributed random variables with cdf $G(t)$.

We will study the random motion of a particle that starts its motion from the origin $\mathbf{0} = (0, 0, \ldots, 0)$ of the space $\mathbb{R}^n$, at time $t = 0$, and continues its motion with a constant velocity $v > 0$ along a direction $\boldsymbol{\eta}_0$ in $\mathbb{R}^n$. At instant τ_1, the particle changes its direction to $\boldsymbol{\eta}_1$ and continues its motion with a velocity $v > 0$ along $\boldsymbol{\eta}_1$. At instant τ_2, the particle changes its direction to $\boldsymbol{\eta}_2$ and continues its motion with a velocity $v > 0$ along $\boldsymbol{\eta}_2$, and so on. We assumed that $\boldsymbol{\eta}_i$, $i = 0, 1, 2, \ldots$ are independent and identically distributed random n-dimensional vectors uniformly distributed on the unit sphere $\Omega_1^{(n-1)} = \{(x_1, x_2, \ldots, x_n) : x_1^2 + x_2^2 + \ldots + x_n^2 = 1\}$.

Denote by $\mathbf{x}(t)$, $t \geq 0$, the particle position at time t. We have:

$$\mathbf{x}(t) = v \sum_{j=1}^{\nu(t)} \boldsymbol{\eta}_{j-1} \theta_{j-1} + v\, \boldsymbol{\eta}_{\nu(t)} \left(t - \tau_{\nu(t)}\right).$$

As we know, the characteristic function $H(t, \alpha) = \mathrm{E}\left[\exp\{i(\boldsymbol{\alpha}, \mathbf{x}(t))\}\right]$, $\boldsymbol{\alpha} = (\alpha_1, \alpha_2, \ldots, \alpha_n)$, $\alpha = |\boldsymbol{\alpha}| = \sqrt{\alpha_1^2 + \alpha_2^2 + \ldots + \alpha_n^2}$ satisfies the following renewal equation (Pogorui and Rodríguez-Dagnino 2011, 2013):

$$H(t, \alpha) = (1 - G(t))\, \varphi(t) + \int_0^t g(u)\, \varphi(u) H(t-u, \alpha)\, du,$$

where $\varphi(t, \alpha) = 2^{\frac{n-2}{2}} \Gamma\left(\frac{n}{2}\right) \frac{J_{\frac{n-2}{2}}(\alpha t v)}{(\alpha t v)^{\frac{n-2}{2}}}$, $G(t)$ is the (m, λ)-Erlang distribution function, and $g(t) = \frac{dG(t)}{dt} = \frac{\lambda^m}{(m-1)!} t^{m-1} e^{-\lambda t} I_{\{t \geq 0\}}$ is the corresponding pdf.

In the following, we will omit α in $H(t, \alpha)$ and $\varphi(t, \alpha)$ to avoid cumbersome formulas, i.e. $H(t) = H(t, \alpha)$, $\varphi(t) = \varphi(t, \alpha)$.

Denote by $f_n(t, \mathbf{x})$ the generalized pdf of the particle position at point $\mathbf{x}$ and time t. It is well known that the function $f_n(t, \mathbf{x})$ is radial, i.e. it only depends on $x = |\mathbf{x}| =$

$\sqrt{x_1^2 + x_2^2 + \ldots + x_n^2}$ Pogorui and Rodríguez-Dagnino (2011, 2013); Orsingher and De Gregorio (2007). Thus, $f_n(t,x) = f_n(t,\mathbf{x}) = \mathcal{F}^{-1}(H(t),\alpha)$, where $\mathcal{F}^{-1}$ is the n-dimensional inverse Fourier transform with respect to α. We will show that for an n-dimensional space $\mathbb{R}^n$ ($n \geq 2$), in the case where $G(t)$ is an $(n-1,\lambda)$-Erlang distribution, the generalized pdf $f_n(t,x)$ is the solution of a telegraph-type equation of order n. We will do this for important cases when n is equal to two, three and five dimensions. The case $n = 4$ will be published elsewhere.

Hence, for $g(t) = \frac{\lambda^{n-1}}{\Gamma(n-1)} t^{n-2} e^{-\lambda t}$, we have

$$H(t) = \left(\sum_{k=0}^{n-2} \frac{(\lambda t)^k e^{-\lambda t}}{k!}\right) 2^{\frac{n-2}{2}} \Gamma\left(\frac{n}{2}\right) \frac{J_{\frac{n-2}{2}}(\alpha t v)}{(\alpha t v)^{\frac{n-2}{2}}} + \frac{\lambda^{n-1} 2^{\frac{n-2}{2}} \Gamma\left(\frac{n}{2}\right)}{\Gamma(n-1)(\alpha v)^{\frac{n-2}{2}}}$$

$$\times \int_0^t (t-u)^{\frac{n-2}{2}} e^{-\lambda(t-u)} J_{\frac{n-2}{2}}(\alpha(t-u)v) H(u)\, du. \qquad [1.51]$$

It is well known that the function $f_n(t,x) = f_n(t,\mathbf{x}) = \mathcal{F}^{-1}(H(t),\alpha)$, where $\mathcal{F}^{-1}$ is the n-dimensional inverse Fourier transform with respect to α, has the following form (Pogorui and Rodríguez-Dagnino 2011, 2013; Orsingher and De Gregorio 2007):

$$f_n(t,\mathbf{x}) = c_n(t,\mathbf{x}) + s_n(t,\mathbf{x}).$$

In this expression, $s_n(t,\mathbf{x})$ represents the singular part and $c_n(t,\mathbf{x})$ is the continuous part of the generalized pdf $f_n(t,\mathbf{x})$. We should note that the generalized pdf $f_n(t,\mathbf{x})$ and functions $s_n(t,\mathbf{x})$, $c_n(t,\mathbf{x})$ are radial, i.e. $f_n(t,\mathbf{x}) = f_n(t,x)$, $s_n(t,\mathbf{x}) = s_n(t,x)$, $c_n(t,\mathbf{x}) = c_n(t,x)$, where $x = |\mathbf{x}|$. The singular term $s_n(t,\mathbf{x})$ corresponds to event $\tau_1 > t$, which has probability $\left(\sum_{k=0}^{n-2} \frac{(\lambda t)^k e^{-\lambda t}}{k!}\right)$, and reflects the fact that the particle has not yet become "smeared out" by the random velocity alternation.

We define $H_c(t) = H_c(t,\alpha) = \int_{\mathbb{R}^n} e^{i(\alpha,\mathbf{x})} c_n(t,\mathbf{x})\, d\mathbf{x}$, $H_s(t) = H_s(t,\alpha) = \int_{\mathbb{R}^n} e^{i(\alpha,\mathbf{x})} s_n(t,\mathbf{x})\, d\mathbf{x}$, hence $H(t) = H_c(t) + H_s(t)$. The characteristic function for the singular part is

$$H_s(t) = \left(\sum_{k=0}^{n-2} \frac{(\lambda t)^k e^{-\lambda t}}{k!}\right) 2^{\frac{n-2}{2}} \Gamma\left(\frac{n}{2}\right) \frac{J_{\frac{n-2}{2}}(\alpha t v)}{(\alpha t v)^{\frac{n-2}{2}}}.$$

Thus, from equation [1.51] we have

$$H_c(t) + H_s(t) = \left(\sum_{k=0}^{n-2} \frac{(\lambda t)^k e^{-\lambda t}}{k!}\right) 2^{\frac{n-2}{2}} \Gamma\left(\frac{n}{2}\right) \frac{J_{\frac{n-2}{2}}(\alpha t v)}{(\alpha t v)^{\frac{n-2}{2}}}$$

$$+\frac{\lambda^{n-1}2^{\frac{n-2}{2}}\Gamma\left(\frac{n}{2}\right)}{\Gamma\left(n-1\right)\left(\alpha v\right)^{\frac{n-2}{2}}}$$

$$\times\int_0^t\left(t-u\right)^{\frac{n-2}{2}}e^{-\lambda(t-u)}J_{\frac{n-2}{2}}\left(\alpha v\left(t-u\right)\right)\left(H_c\left(u\right)+H_s\left(u\right)\right)du.$$

Furthermore,

$$H_c(t)=\frac{\lambda^{n-1}2^{n-2}\left(\Gamma\left(\frac{n}{2}\right)\right)^2}{\Gamma\left(n-1\right)\left(\alpha v\right)^{n-2}}\int_0^t\left(t-u\right)^{\frac{n-2}{2}}e^{-\lambda(t-u)}J_{\frac{n-2}{2}}\left(\alpha v\left(t-u\right)\right)$$

$$\times\left(\sum_{k=0}^{n-2}\frac{(\lambda u)^k e^{-\lambda u}}{k!}\right)\frac{J_{\frac{n-2}{2}}\left(\alpha uv\right)}{u^{\frac{n-2}{2}}}du$$

$$+\frac{\lambda^{n-1}2^{\frac{n-2}{2}}\Gamma\left(\frac{n}{2}\right)}{\Gamma\left(n-1\right)\left(\alpha v\right)^{\frac{n-2}{2}}}\int_0^t\left(t-u\right)^{\frac{n-2}{2}}e^{-\lambda(t-u)}J_{\frac{n-2}{2}}\left(\alpha v\left(t-u\right)\right)H_c\left(u\right)du.$$

Now, by setting $H_c(t)e^{\lambda t}=h_c(t)$, we have

$$h_c(t)=\frac{\lambda^{n-1}2^{n-2}\left(\Gamma\left(\frac{n}{2}\right)\right)^2}{\Gamma\left(n-1\right)\left(\alpha v\right)^{n-2}}\int_0^t\left(t-u\right)^{\frac{n-2}{2}}J_{\frac{n-2}{2}}\left(\alpha v\left(t-u\right)\right)\frac{J_{\frac{n-2}{2}}\left(\alpha uv\right)}{u^{\frac{n-2}{2}}}$$

$$\times\left(\sum_{k=0}^{n-2}\frac{(\lambda u)^k}{k!}\right)du$$

$$+\frac{\lambda^{n-1}2^{\frac{n-2}{2}}\Gamma\left(\frac{n}{2}\right)}{\Gamma\left(n-1\right)\left(\alpha v\right)^{\frac{n-2}{2}}}\int_0^t\left(t-u\right)^{\frac{n-2}{2}}J_{\frac{n-2}{2}}\left(\alpha v\left(t-u\right)\right)h_c\left(u\right)du. \qquad [1.52]$$

Suppose that $S_n\left(R\right)$ is a sphere with a radius R in $\mathbb{R}^n$ and f is a continuous function defined over $S_n\left(R\right)$. Let us consider the generalized function $f\delta_{S_n(R)}$, acting according to the following rule:

$$\left(f\delta_{S_n(R)},\varphi\right)=\int_{S_n(R)}f\left(\mathbf{x}\right)\varphi\left(\mathbf{x}\right)\mu\left(d\mathbf{x}\right),$$

where φ is an infinitely differentiable function with compact support and μ is the Lebesgue measure in $\mathbb{R}^n$. The function is called a simple layer on the spherical surface $S_n\left(R\right)$ (Vladimirov 1996).

1.4.2. *Two-dimensional case*

For $n = 2$, equation [1.51] can be written in the following form:

$$H(t) = e^{-\lambda t} J_0 (\alpha t v) + \lambda \int_0^t e^{-\lambda(t-u)} J_0 (\alpha (t-u) v) H (u) du$$

and equation [1.52] has the form

$$h_c(t) = \lambda \int_0^t J_0 (\alpha v (t-u)) J_0 (\alpha v u) \, du$$

$$+\lambda \int_0^t J_0 (\alpha v (t-u)) h_c (u) \, du. \qquad [1.53]$$

THEOREM 1.5.– The continuous part $c_2 (t, \mathbf{x})$ of the generalized pdf $f_2 (t, \mathbf{x})$ satisfies the following equation for $0 \leq x \leq vt$:

$$\frac{\partial^2 c_2 (t, \mathbf{x})}{\partial t^2} + 2\lambda \frac{\partial}{\partial t} c_2 (t, \mathbf{x}) - v^2 \left(\frac{\partial^2 c_2 (t, \mathbf{x})}{\partial x_1^2} + \frac{\partial^2 c_2 (t, \mathbf{x})}{\partial x_2^2} \right)$$

$$= \lambda^2 e^{-\lambda t} \frac{\delta_{S_2(vt)}}{2\pi v t}, \qquad [1.54]$$

with the initial conditions $c_2(0) = 0$ and $\left. \frac{\partial c_2(t)}{\partial t} \right|_{t=0} = \lambda \delta(x)$.

DEMONSTRATION.– Let us consider the first two derivatives of $h_c(t)$

$$\frac{\partial}{\partial t} h_c(t) = \lambda J_0 (\alpha v t) - \lambda \alpha v \int_0^t J_1 (\alpha v (t-u)) J_0 (\alpha v u) \, du$$

$$+\lambda h_c(t) - \lambda \alpha v \int_0^t J_1 (\alpha v (t-u)) h_c (u) \, du \qquad [1.55]$$

and

$$\frac{\partial^2}{\partial t^2} h_c(t) = \lambda \frac{\partial}{\partial t} (J_0 (\alpha v t)) - \lambda (\alpha v)^2 \int_0^t J_0 (\alpha v (t-u)) J_0 (\alpha v u) \, du$$

$$+\lambda (\alpha v)^2 \int_0^t \frac{J_1 (\alpha v (t-u))}{\alpha v (t-u)} J_0 (\alpha v u) \, du + \lambda \frac{d}{dt} h_c(t)$$

$$-\lambda (\alpha v)^2 \int_0^t J_0 (\alpha v (t-u)) h_c (u) \, du$$

$$+\lambda (\alpha v)^2 \int_0^t \frac{J_1 (\alpha v (t-u))}{\alpha v (t-u)} h_c (u) \, du. \qquad [1.56]$$

The last integral in equation [1.56] can be written in a more convenient form by doing some equivalent transformations

$$\int_0^t \frac{J_1(\alpha v(t-u))}{\alpha v(t-u)} h_c(u)\,du$$

$$= \lambda \int_0^t \frac{J_1(\alpha v(t-u))}{\alpha v(t-u)} \int_0^u J_0(\alpha v(u-s)) J_0(\alpha v s)\,ds\,du$$

$$+\lambda \int_0^t \frac{J_1(\alpha v(t-u))}{\alpha v(t-u)} \int_0^u J_0(\alpha v(u-s)) h_c(s)\,ds\,du$$

$$= \lambda \int_0^t \int_0^{t-s} \frac{J_1(\alpha v(t-s-u))}{\alpha v(t-s-u)} J_0(\alpha v u)\,du\, J_0(\alpha v s)\,ds$$

$$+\lambda \int_0^t \int_0^{t-s} \frac{J_1(\alpha v(t-s-u))}{\alpha v(t-s-u)} J_0(\alpha v u)\,du\, h_c(s)\,ds$$

$$= \frac{\lambda}{\alpha v} \int_0^t J_1(\alpha v(t-s))\, J_0(\alpha v s)\,ds + \frac{\lambda}{\alpha v} \int_0^t J_1(\alpha v(t-s))\, h_c(s)\,ds,$$

where we used the formula 6.533 in Gradshteyn and Ryzhik (1980); we repeat that formula here for easy reference,

$$\int_0^x \frac{J_1(x-u)}{x-u} J_0(u)\,du = J_1(x).$$

Thus, combining equations [1.53]–[1.56], and taking into account that $\lambda \frac{\partial}{\partial t} J_0(\alpha v t) + \lambda \alpha v J_1(\alpha v t) = 0$, we obtain

$$\frac{\partial^2}{\partial t^2} h_c(t) = \left\{ \lambda \frac{\partial}{\partial t} (J_0(\alpha v t)) + \lambda \alpha v J_1(\alpha v t) \right\}$$

$$-\lambda (\alpha v)^2 \int_0^t J_0(\alpha v(t-u)) J_0(\alpha v u)\,du$$

$$-\lambda (\alpha v)^2 \int_0^t J_0(\alpha v(t-u)) h_c(u)\,du + \lambda \frac{\partial}{\partial t} h_c(t)$$

$$+\lambda^2 \alpha v \int_0^t J_1(\alpha v(t-u))\, J_0(\alpha v u)\,du + \lambda^2 \alpha v \int_0^t J_1(\alpha v(t-u))\, h_c(u)\,du$$

$$= -(\alpha v)^2 h_c(t) + \lambda \frac{\partial}{\partial t} h_c(t) - \lambda \left(\frac{\partial}{\partial t} h_c(t) - \lambda J_0(\alpha v t) - \lambda h_c(t) \right).$$

Therefore, we have the equivalent differential equation

$$\frac{\partial^2}{\partial t^2} h_c(t) + (\alpha v)^2 h_c(t) - \lambda^2 h_c(t) = \lambda^2 J_0(\alpha v t). \qquad [1.57]$$

with the initial conditions $h_c(0) = 0$ and $\left.\frac{\partial h_c(t)}{\partial t}\right|_{t=0} = \lambda$, which can be obtained directly from equation [1.53].

It is well known (it can be easily obtained from formula 4.644 of Gradshteyn and Ryzhik 1980) that the function $J_0(\alpha vt)$ is the two-dimensional Fourier transform with respect to $\mathbf{x} = (x_1, x_2)$ of the function $\frac{\delta_{S_2(vt)}}{2\pi vt}$, where $\delta_{S_2(vt)}$ is a simple layer for a circle with radius vt. Thus, $\mathcal{F}^{-1}(J_0(\alpha vt)) = \frac{\delta_{S_2(vt)}}{2\pi vt}$. Denote by $g(t, \mathbf{x}) = \mathcal{F}^{-1}(h_c(t))$ the two-dimensional inverse Fourier transform of $h_c(t)$ with respect to α. Here, we use the notation $g(t, \mathbf{x})$, where g is a radial function, and $g(t, \mathbf{x}) = g(t, \mathbf{y})$ for all $\mathbf{x}$, $\mathbf{y} \in \mathbb{R}^2$ such that $|\mathbf{x}| = |\mathbf{y}|$.

We have the following two-dimensional inverse Fourier transform:

$$\mathcal{F}^{-1}\left(\alpha^2 h_c(t), \alpha\right) = -\sum_{i=1}^{2} \frac{\partial^2 g(t, \mathbf{x})}{\partial x_i^2}.$$

Here, we used the fact that $\mathcal{F}^{-1}\left(\alpha_i^2 h_c(t), \alpha\right) = -\frac{\partial^2 g(t,\mathbf{x})}{\partial x_i^2}$. Passing to the two-dimensional inverse Fourier transform in equation [1.57], for $0 \le x \le vt$ we can obtain

$$\frac{\partial^2}{\partial t^2} g(t, \mathbf{x}) + (\alpha v)^2 g(t, \mathbf{x}) - \lambda^2 g(t, \mathbf{x}) = \lambda^2 \frac{\delta_{S_2(vt)}}{2\pi vt}. \qquad [1.58]$$

Let us denote by $c_2(t, \mathbf{x}) = \mathcal{F}^{-1}(H_c(t), \alpha)$, the absolute continuous part of the generalized pdf of the particle's position at time t.

Since $e^{\lambda t} c_2(t, \mathbf{x}) = g(t, \mathbf{x})$, from equation [1.58] we obtain equation [1.54], with the initial conditions $c_2(0) = 0$ and $\left.\frac{\partial c_2(t)}{\partial t}\right|_{t=0} = \lambda\delta(x)$, which can be obtained directly from the initial conditions of equation [1.57].

The fact that $c_2(t, \mathbf{x})$ is a fundamental solution of the two-dimensional Goldstein–Kac equation, has been found in literature by using other methods. However, in our case, equation [1.54] differs from the corresponding similar published equation, and it is a consequence of specifying the expression on the right side of the equation and the initial conditions.

At first glance, it may seem strange that on the right side of equation [1.54] we have a singular function for the continuous part $c_2(t, \mathbf{x})$. But there is no contradiction if we consider that the continuous part $c_2(t, \mathbf{x})$ goes to infinity as $|x| \to vt$, and it has the so-called explosion effect for just the continuous part of the generalized density (Pogorui and Rodríguez-Dagnino 2011, and the references therein).

Thus, the generalized pdf of the particle position is given by $f(t, \mathbf{x}) = c_2(t, \mathbf{x}) + s_2(t, \mathbf{x})$, where $c_2(t, \mathbf{x})$ is the solution of the two-dimensional Goldstein–Kac telegraph equation [1.54] and $s_2(t, \mathbf{x}) = e^{-\lambda t}\delta(vt - x)$.

REMARK 1.3.– *We will apply the basic method developed for $n = 2$, to the cases $n=3$ and $n = 5$ in the following sections. We should remark that most of the published works in this area have solved the cases with even dimensions, and the cases of odd dimensions remain as elusive cases. We will show, by using our method, that it is possible to find the governing telegraph-type partial differential equations for these odd dimensions, and we recall the corresponding solution in the three-dimensional case.*

1.4.3. *Three-dimensional case*

In this case for a $(2, \lambda)$-Erlang pdf (i.e. $g(t) = \lambda^2 te^{-\lambda t},\ t \geq 0$), the characteristic function $H(t)$ satisfies the following integral equation:

$$H(t) = (1 + \lambda t)\, e^{-\lambda t} \frac{\sin(\alpha vt)}{\alpha vt}$$

$$+\frac{\lambda^2}{\alpha v} \int_0^t e^{-\lambda(t-u)} \sin(\alpha v\,(t-u)) H(u) du. \qquad [1.59]$$

The generalized pdf $f_3\,(t, x) = \mathcal{F}^{-1}\,(H(t), \alpha)$, where $x = \sqrt{x_1^2 + x_2^2 + x_3^2} = |\mathbf{x}|$, $\mathcal{F}^{-1}$ is the three-dimensional Fourier transform with respect to α, can be written as follows:

$$f_3\,(t, x) = s_3\,(t, x) + c_3\,(t, x)\,,$$

where $s_3\,(t, x)$ is the singular part and $c_3\,(t, x)$ is the continuous part of the generalized pdf $f_3\,(t, x)$.

Let us define $\Lambda_* = \sqrt{(tv)^2 - x^2}$, then our main result in this section is established in the following basic theorem.

THEOREM 1.6.– For $0 \leq x \leq vt$, the generalized pdf $f_3\,(t, x)$ is given by

$$f_3\,(t, x) = (1 + \lambda t)\, e^{-\lambda t} \frac{\delta_{S_3(vt)}}{4\pi (vt)^2} + \frac{\lambda^2 e^{-\lambda t}}{4\pi v^2} \frac{I_1\left(\frac{\lambda}{v}\Lambda_*\right)}{\Lambda_*}\, \Theta\,(tv - x)$$

$$+\frac{t\lambda e^{-\lambda t}\left(\lambda \Lambda_* I_0\left(\frac{\lambda}{v}\Lambda_*\right) - 2vI_1\left(\frac{\lambda}{v}\Lambda_*\right)\right)}{4\pi v\Lambda_*^3}\,\Theta\,(tv - x) + r\,(t, \mathbf{x})\,, \qquad [1.60]$$

where $\Theta\,(tv - x)$ is the Heaviside function and the reminder $r\,(t, \mathbf{x})$ is non-negative, $r\,(t, \mathbf{x}) = 0$ for $tv < x$ and satisfies (see Gradshteyn and Ryzhik 1980, formula 4.642)

$$\int_{|\mathbf{x}| \leq vt} r\,(t, \mathbf{x})\, d\mathbf{x} = 4\pi \int_0^{vt} x^2 r\,(t, x) dx \leq \frac{\lambda^2 t^2}{2} e^{-\lambda t}.$$

DEMONSTRATION.– As shown in Vladimirov (1996), the function $4\pi R\frac{\sin \alpha R}{\alpha}$ is the three-dimensional Fourier transform of the simple layer $\delta_{S_3(R)}$. Hence,

$$s_3(t,x) = \mathcal{F}^{-1}\left((1+\lambda t)\, e^{-\lambda t}\, \frac{\sin(\alpha v t)}{\alpha v t}\right) = (1+\lambda t)\, e^{-\lambda t}\, \frac{\delta_{S_3(vt)}}{4\pi (vt)^2}.$$

It follows from equation [1.59] that

$$h_c(t) = \lambda^2 \int_0^t \frac{\sin(\alpha v (t-u))}{\alpha v} (1+\lambda u) \frac{\sin(\alpha v u)}{\alpha v u} du$$

$$+\frac{\lambda^2}{\alpha v} \int_0^t \sin(\alpha v (t-u))\, h_c(u)\, du, \qquad [1.61]$$

The second derivative of equation [1.61] with respect to t is given by

$$\frac{\partial^2 h_c(t)}{\partial t^2} - \lambda^2 h_c(t) + \lambda^2 \alpha v \int_0^t \sin(\alpha v (t-u))\, h_c(u)\, du$$

$$= (\lambda^2 + \lambda^3 t) \frac{\sin(\alpha v t)}{\alpha v t}$$

$$-\lambda^2 \alpha v \int_0^t \sin(\alpha v (t-u))\ (1+\lambda u) \frac{\sin(\alpha v u)}{u} du. \qquad [1.62]$$

Combining equations [1.61] and [1.62], we obtain a differential equation for $h_c(t)$

$$\frac{\partial^2 h_c(t)}{\partial t^2} - \lambda^2 h_c(t) + \alpha^2 v^2 h_c(t) = (\lambda^2 + \lambda^3 t) \frac{\sin(\alpha v t)}{\alpha v t} \qquad [1.63]$$

with the initial conditions $h_c(0) = 0$, $\left.\frac{\partial h_c(t)}{\partial t}\right|_{t=0} = 0$ that can be calculated directly from equation [1.61].

Since a solution of equation [1.63] with the given initial conditions is unique, we can split equation [1.63] into a set of two equations. Namely,

$$\frac{\partial^2 h_c^{(1)}(t)}{\partial t^2} - \lambda^2 h_c^{(1)}(t) + \alpha^2 v^2 h_c^{(1)}(t) = \lambda^2 \frac{\sin(\alpha v t)}{\alpha v t} \qquad [1.64]$$

with the initial conditions $h_c^{(1)}(0) = 0$, $\left.\frac{\partial h_c^{(1)}(t)}{\partial t}\right|_{t=0} = 0$, and

$$\frac{\partial^2 h_c^{(2)}(t)}{\partial t^2} - \lambda^2 h_c^{(2)}(t) + \alpha^2 v^2 h_c^{(2)}(t) = \lambda^3 \frac{\sin(\alpha v t)}{\alpha v} \qquad [1.65]$$

with similar initial conditions $h_c^{(2)}(0) = 0$, $\left.\frac{\partial h_c^{(2)}(t)}{\partial t}\right|_{t=0} = 0$.

The solutions of equations [1.64] and [1.65] with the given initial conditions are also unique. Hence, the solution $h_c(t)$ of equation [1.63] is

$$h_c(t) = h_c^{(1)}(t) + h_c^{(2)}(t).$$

It is easily verified that by solving equation [1.64], we can obtain

$$\begin{aligned} h_c^{(1)}(t) &= \frac{\lambda^2}{2\alpha v \Lambda_o}\Big(\cos(t\Lambda_o)\,\mathrm{Ci}\,(t(\alpha v - \Lambda_o)) \\ &\quad - \cos(t\Lambda_o)\,\mathrm{Ci}\,(t(\Lambda_o + \alpha v)) + \sin(t\Lambda_o)\,\mathrm{Si}\,(t(\Lambda_o - \alpha v)) \\ &\quad - \sin(t\Lambda_o)\,\mathrm{Si}\,(t(\Lambda_o - \alpha v)) + \cos(t\Lambda_o)\left(\ln(\alpha v + \Lambda_o) - \ln(\alpha v - \Lambda_o)\right)\Big), \end{aligned}$$

where $\Lambda_o = \sqrt{\alpha^2 v^2 - \lambda^2}$, $\mathrm{Ci}(\cdot)$ is the cosine integral, and $\mathrm{Si}(\cdot)$ is the sine integral.

Then, by solving equation [1.65], we obtain

$$h_c^{(2)}(t) = \frac{\lambda \sin(t\Lambda_o)}{\Lambda_o} - \frac{\lambda \sin(\alpha v t)}{\alpha v}.$$

By using formula 2.15.10 (4) of Prudnikov *et al.* (1986) and the Hankel transform of order $\frac{1}{2}$, after some elementary mathematical simplifications, we obtain the three-dimensional inverse Fourier transform of $h_c^{(2)}(t)$ for $x \le vt$

$$\begin{aligned} g^{(2)}(t,\mathbf{x}) &= \mathcal{F}^{-1}\left(h_c^{(2)}(t), \alpha\right) = \lambda \mathcal{F}^{-1}\left(\frac{\sin(t\Lambda_o)}{\Lambda_o} - \frac{\sin(\alpha v t)}{\alpha v}, \alpha\right) \\ &= \frac{\lambda^2}{4\pi v^2}\frac{I_1\left(\frac{\lambda}{v}\Lambda_*\right)}{\Lambda_*}\Theta(tv - x). \end{aligned}$$

We should note that this result coincides with the result obtained in Garra and Orsingher (2014, equation 4.1). However, it represents only a partial result in our analysis and to obtain $c_3(t,x)$, we should calculate $g^{(1)}(t,\mathbf{x}) = \mathcal{F}^{-1}\left(h_c^{(1)}(t), \alpha\right)$. Unfortunately, we cannot obtain a closed-form solution for this expression. A good approximated solution can be obtained by taking into account the inverse three-dimensional Fourier transform $\mathcal{F}^{-1}\left(\alpha^2 h_c^{(1)}, \alpha\right) = -\sum_{i=1}^{3} \frac{\partial^2 g^{(1)}(t,\mathbf{x})}{\partial x_i^2}$, and passing to the inverse three-dimensional Fourier transform of equation [1.63], it is easily seen that for $x < vt$

$$\frac{\partial^2 g^{(1)}(t,\mathbf{x})}{\partial t^2} - \lambda^2 g^{(1)}(t,\mathbf{x}) - v^2 \sum_{i=1}^{3} \frac{\partial^2 g^{(1)}(t,\mathbf{x})}{\partial x_i^2} = 0. \qquad [1.66]$$

Here we used the fact that $\mathcal{F}^{-1}\left(\frac{\sin(\alpha v t)}{\alpha v t}, \alpha\right) = \frac{\delta_{S(vt)}}{4\pi(vt)^2} = 0$ for $x < vt$, where $\delta_{S(vt)}$ is a three-dimensional simple layer.

It is straightforward to verify that

$$\tilde{g}^{(1)}(t,\mathbf{x}) = \frac{\lambda}{4\pi v^2}\frac{\partial}{\partial t}\frac{I_1\left(\frac{\lambda}{v}\Lambda_*\right)}{\Lambda_*} = \frac{t\lambda\left(\lambda\Lambda_* I_0\left(\frac{\lambda}{v}\Lambda_*\right) - 2vI_1\left(\frac{\lambda}{v}\Lambda_*\right)\right)}{4\pi v\Lambda_*^3}$$

is a solution of equation [1.66], for $x < vt$.

Then, since

$$\lim_{\alpha\to 0} h_c^{(2)}(t) = \lim_{\alpha\to 0}\left(\frac{\lambda\sin(t\Lambda_o)}{\Lambda_o} - \frac{\lambda\sin(\alpha vt)}{\alpha v}\right) = \sinh(\lambda t) - \lambda t,$$

and by using (Gradshteyn and Ryzhik 1980, formula 4.642) we have

$$\int_{|\mathbf{x}|\le vt} g^{(2)}(t,\mathbf{x})\,d\mathbf{x} = 4\pi\int_0^{vt} x^2 g^{(2)}(t,x)\,dx$$

$$= \frac{\lambda^2}{v^2}\int_0^{tv} x^2\frac{I_1\left(\frac{\lambda}{v}\Lambda_*\right)}{\Lambda_*}\,dx = \sinh(\lambda t) - \lambda t.$$

Hence,

$$\frac{\lambda}{v^2}\int_0^{tv} x^2\frac{\partial}{\partial t}\frac{I_1\left(\frac{\lambda}{v}\Lambda_*\right)}{\Lambda_*}\,dx = \cosh(\lambda t) - 1 - \frac{\lambda^2 t^2}{2}.$$

In a similar manner,

$$\int_{|\mathbf{x}|\le vt}\tilde{g}^{(1)}(t,\mathbf{x})\,d\mathbf{x} = \frac{\lambda}{v}\int_0^{tv} x^2\frac{t\left(\lambda\Lambda_* I_0\left(\frac{\lambda}{v}\Lambda_*\right) - 2vI_1\left(\frac{\lambda}{v}\Lambda_*\right)\right)}{\Lambda_*^3}\,dx$$

$$= \cosh(\lambda t) - 1 - \frac{\lambda^2 t^2}{2}.$$

On the other hand, it can be easily verified that

$$\lim_{\alpha\to 0} h_c^{(1)}(t) = \lim_{\alpha\to 0}\frac{\lambda^2}{2\alpha v\Lambda_o}\Bigg(\cos(t\Lambda_o)\,\mathrm{Ci}\left(t\left(\alpha v - \Lambda_o\right)\right)$$

$$-\cos(t\Lambda_o)\,\mathrm{Ci}\left(t\left(\Lambda_o + \alpha v\right)\right) + \sin(t\Lambda_o)\,\mathrm{Si}\left(t\left(\Lambda_o - \alpha v\right)\right)$$

$$-\sin(t\Lambda_o)\,\mathrm{Si}\left(t\left(\Lambda_o - \alpha v\right)\right) + \cos(t\Lambda_o)\left(\ln(\alpha v + \Lambda_o) - \ln(\alpha v - \Lambda_o)\right)\Bigg)$$

$$= \cosh(\lambda t) - 1.$$

Hence, taking into account (Gradshteyn and Ryzhik 1980, formula 4.642), we have

$$\int_{|\mathbf{x}|<vt} g^{(1)}(t,\mathbf{x})\, d\mathbf{x} = 4\pi \int_0^{vt} x^2 g^{(1)}(t,x)\, dx = \cosh(\lambda t) - 1.$$

By defining $r(t,x) = g^{(1)}(t,x) - \tilde{g}^{(1)}(t,x)$ and considering that

$$c_3(t,\mathbf{x}) = \mathcal{F}^{-1}(H_c(t),\alpha) = \left(g^{(1)}(t,\mathbf{x}) + g^{(2)}(t,\mathbf{x})\right) e^{-\lambda t}$$

the proof is concluded. □

REMARK 1.4.– *We should mention that* $c_3(t,\mathbf{x})$ *satisfies*

$$\frac{\partial^2 c_3(t,\mathbf{x})}{\partial t^2} + 2\lambda \frac{\partial}{\partial t} c_3(t,\mathbf{x}) - v^2 \sum_{i=1}^{3} \frac{\partial^2 c_3(t,\mathbf{x})}{\partial x_i^2}$$

$$= \left(\lambda^2 + \lambda^3 t\right) e^{-\lambda t} \frac{\delta_{S_3(vt)}}{4\pi (vt)^2}. \qquad [1.67]$$

$$c_3(0,\mathbf{x}) = 0, \quad \left.\frac{\partial c_3(t,\mathbf{x})}{\partial t}\right|_{t=0} = 0.$$

We have a singular function on the right side of equation [1.67] *for the continuous part* $c_3(t,\mathbf{x})$ *since it goes to infinity as* $|x| \to vt$, *and it has the explosion effect just as well as in the two-dimensional case for* $c_2(t,\mathbf{x})$.

Let us assume $g(t,\mathbf{x}) = \mathcal{F}^{-1}(h_c(t),\alpha)$, where $h_c(t)$ is a solution of equation [1.63]. Since $e^{\lambda t} c_3(t,\mathbf{x}) = g(t,\mathbf{x})$, and in passing equation [1.63] to the three-dimensional inverse Fourier transform, we obtain equation [1.67].

The generalized pdf $f_3(t,x)$ found in equation [1.60] depends on λ and v, and we illustrate its behavior in Figure 1.1 for $\lambda = 2$ and $v = 3$.

Similarly, we plot equation [4.1] of Garra and Orsingher (2014) in Figure 1.2.

In order to do more comparisons, we will include the singular part to the generalized pdf $f(t,\mathbf{x}) = f(t,x)$ found in Garra and Orsingher (2014), i.e.

$$f(t,x) = (1+\lambda t)\, e^{-\lambda t} \frac{\delta_{S(vt)}}{4\pi (vt)^2} + \frac{\lambda^2 e^{-\lambda t}}{4\pi v^2} \frac{I_1\left(\frac{\lambda}{v}\Lambda_*\right)}{\Lambda_*} \Theta(tv - x). \qquad [1.68]$$

We were not able to find a closed-form expression for $r(t,x)$, however, we are now including the singular part of $f_3(t,x) - r(t,x)$. The comparison of the integrals on the whole range of equations [1.60] and [1.68] are shown in Figures 1.3–1.6 for different combinations of λ and v.

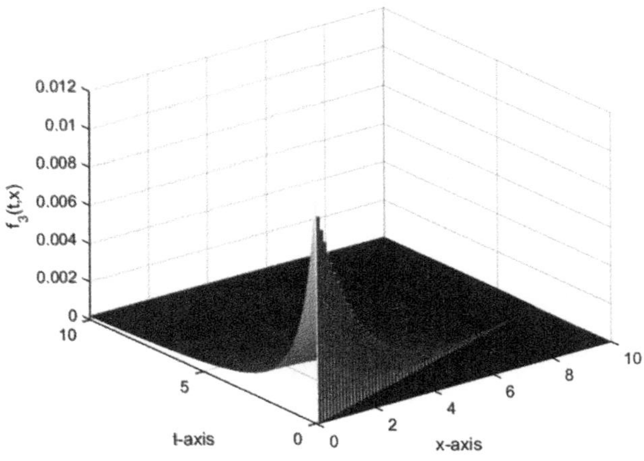

Figure 1.1. *Approximated probability density function $f_3(t,x) - r(t,x)$ for $\lambda = 2$ and $v = 3$. The singular term is not included in this plot. For a color version of this figure, see www.iste.co.uk/pogorui/random2.zip*

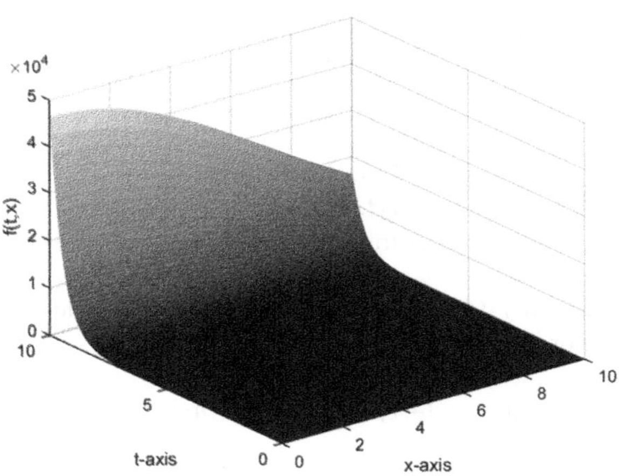

Figure 1.2. *$f(t,x)$ for $\lambda = 2$ and $v = 3$, according to Garra and Orsingher (2014). For a color version of this figure, see www.iste.co.uk/pogorui/random2.zip*

The results in these plots show that $f_3(t,x)$ is a pdf and the contribution of $r(t,x)$ is not significant for t large, since the integral approaches 1 very quickly. We should mention that the result in Garra and Orsingher (2014) seems to just concern the probability of finding the particle after an odd number of direction changes. So, it is not a true pdf since "half" of the time particle moves after odd and "half" after even direction changes. It is not surprising that $f(t,x)$ goes to 1/2 as t is large.

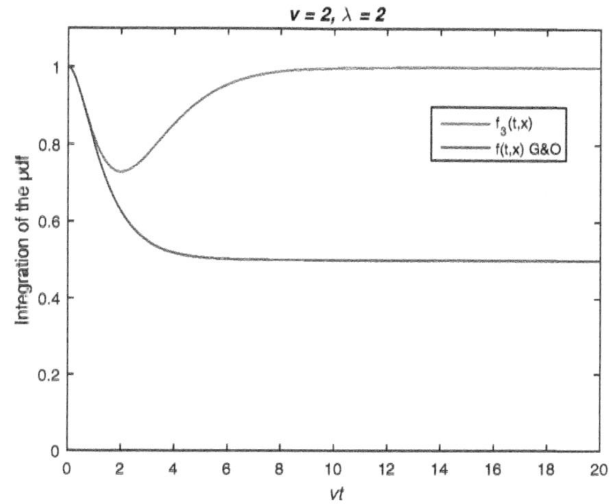

Figure 1.3. *Integration of $f_3(t,x) - r(t,x)$ and $f(t,x)$ for $\lambda = 2$ and $v = 2$. The singular part is included in both cases. For a color version of this figure, see www.iste.co.uk/pogorui/random2.zip*

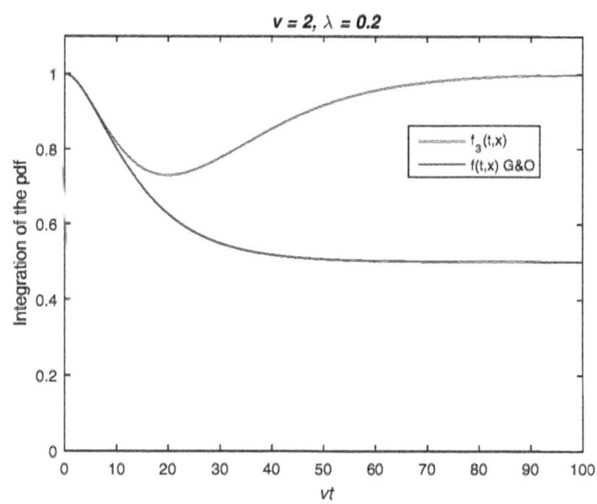

Figure 1.4. *Integration of $f_3(t,x) - r(t,x)$ and $f(t,x)$ for $\lambda = 0.2$ and $v = 2$. The singular part is included in both cases. For a color version of this figure, see www.iste.co.uk/pogorui/random2.zip*

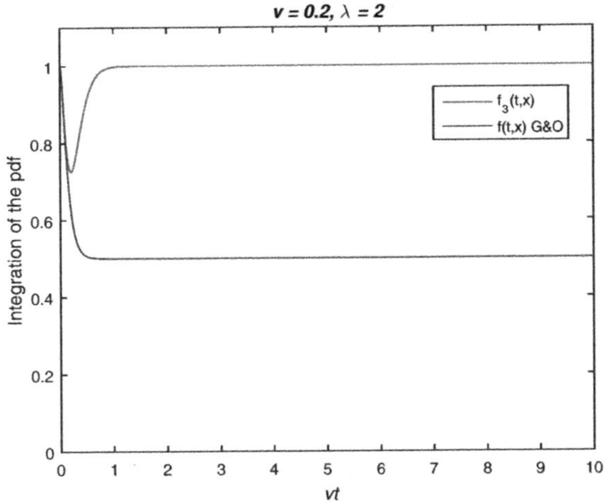

Figure 1.5. *Integration of* $f_3(t,x) - r(t,x)$ *and* $f(t,x)$ *for* $\lambda = 2$ *and* $v = 0.2$*. The singular part is included in both cases. For a color version of this figure, see www.iste.co.uk/pogorui/random2.zip*

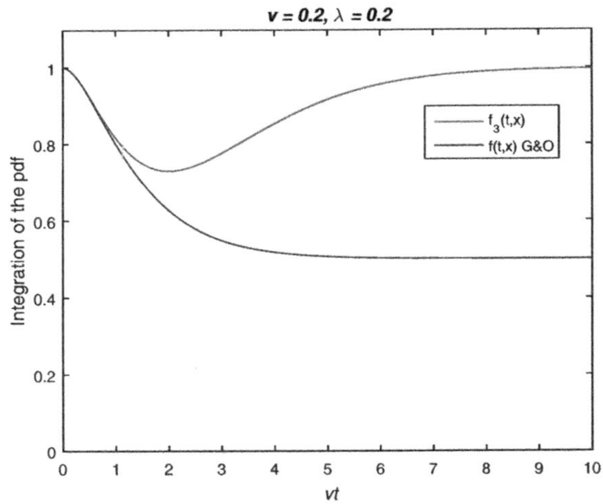

Figure 1.6. *Integration of* $f_3(t,x)$ *and* $f(t,x)$ *for* $\lambda = 0.2$ *and* $v = 0.2$*. The singular part is included in both cases. For a color version of this figure, see www.iste.co.uk/pogorui/random2.zip*

We should mention that equation 7.4.18 in Morse and Feshbach (1953) was proposed as a model for the transmission of heat in a gas, where the velocity of the propagation of the heat is included, or the velocity of the propagation of a disturbance in a gas. A similar reasoning was considered in Tautz and Lerche (2016) for modeling the transport of solar energetic particles, where the authors emphasize the fact that the telegraph-type equation has the potential to make a difference between early ballistic motion and later diffusive transport (Kac's limit). We should emphasize that the underlying stochastic process for this random flight is governed by 2-Erlang-distributed sojourn times. So, in this work we found that the non-homogeneous three-dimensional Goldstein–Kac telegraph equation is the partial differential equation for modeling this random flight.

1.4.4. *Five-dimensional case*

In this case, we have

$$\varphi(t,\alpha) = 2^{\frac{3}{2}}\Gamma\left(\frac{5}{2}\right)\frac{J_{\frac{3}{2}}(\alpha vt)}{(\alpha vt)^{\frac{3}{2}}} = 3\,\frac{\sin(\alpha vt) - \alpha vt\cos(\alpha vt)}{(\alpha vt)^3},$$

where

$$\alpha = \sqrt{\alpha_1^2+\alpha_2^2+\alpha_3^2+\alpha_4^2+\alpha_5^2}.$$

Considering the case of a $(4,\lambda)$-Erlang pdf, i.e. $g(t) = \frac{1}{6}\lambda^4 t^3 e^{-\lambda t}\mathbb{I}_{\{t\geq 0\}}$, we obtain

$$\begin{aligned}H(t) &= \left(\sum_{k=0}^{3}\frac{(\lambda t)^k}{k!}\right)e^{-\lambda t}\,3\,\frac{\sin(\alpha vt) - \alpha vt\cos(\alpha vt)}{(\alpha vt)^3}\\ &+\frac{\lambda^4}{2(\alpha v)^3}\int_0^t e^{-\lambda(t-u)}\left\{\sin(\alpha v(t-u)) - \alpha v(t-u)\cos(\alpha v(t-u))\right\}\\ &\quad\times H(u)\,du. \end{aligned} \qquad [1.69]$$

THEOREM 1.7.– The continuous part $c_5(t,\mathbf{x})$ of the generalized pdf $f_5(t,\mathbf{x})$ satisfies the following equation for $0 \leq x \leq vt$:

$$\begin{aligned}&\left(\frac{\partial^2}{\partial t^2} + 2\lambda\frac{\partial}{\partial t} - v^2\sum_{i=1}^{5}\frac{\partial^2}{\partial x_i^2} + 2\lambda^2\right)\left(\frac{\partial^2}{\partial t^2} + 2\lambda\frac{\partial}{\partial t} - v^2\sum_{i=1}^{5}\frac{\partial^2}{\partial x_i^2}\right)c_5(t,\mathbf{x})\\ &= \lambda^4 e^{-\lambda t}\left(\sum_{k=0}^{3}\frac{(\lambda t)^k}{k!}\right)\delta_{S_5(vt)}.\end{aligned} \qquad [1.70]$$

DEMONSTRATION.– It is well known that the function $3\,\frac{\sin(\alpha vt) - \alpha vt\cos(\alpha vt)}{(\alpha vt)^3}$ is the five-dimensional Fourier transform with respect to x of the function $\frac{8\delta_{S_5(vt)}}{3\pi^2(vt)^4}$, where

$\mathbf{x} = (x_1, x_2, x_3, x_4, x_5)$, $x = |\mathbf{x}|$, and $S_5(R)$ is the simple layer on the spherical surface, with radius R in $\mathbb{R}^5$. So,

$$s_5(t,x) = \mathcal{F}^{-1}\left(\left(\sum_{k=0}^{3} \frac{(\lambda t)^k}{k!}\right) e^{-\lambda t}\, 3\, \frac{\sin(\alpha vt) - \alpha vt\cos(\alpha vt)}{(\alpha vt)^3}, \alpha\right)$$

$$= \left(\sum_{k=0}^{3} \frac{(\lambda t)^k}{k!}\right) e^{-\lambda t}\, \delta_{S(x,vt)}.$$

From equation [1.69], it follows that

$$h_c(t) = \frac{3\lambda^4}{2(\alpha v)^3}\int_0^t \{\sin(\alpha v(t-u)) - \alpha v(t-u)\cos(\alpha v(t-u))\}$$

$$\times \left(\sum_{k=0}^{3} \frac{(\lambda u)^k}{k!}\right) \frac{\sin(\alpha vu) - \alpha vu\cos(\alpha vu)}{(\alpha vu)^3}\, du$$

$$+\frac{\lambda^4}{2(\alpha v)^3}\int_0^t \{\sin(\alpha v(t-u)) - \alpha v(t-u)\cos(\alpha v(t-u))\}$$

$$\times h_c(u)\, du. \qquad [1.71]$$

Consider the following derivatives:

$$\frac{\partial^2}{\partial t^2}\int_0^t \{\sin(\alpha v(t-u)) - \alpha v(t-u)\cos(\alpha v(t-u))\}\, h_c(u)\, du$$

$$= \int_0^t \left\{(\alpha v)^2 \sin(\alpha v(t-u)) + (\alpha v)^3 (t-u)\cos(\alpha v(t-u))\right\}$$

$$\times h_c(u)\, du, \qquad [1.72]$$

and

$$\frac{\partial^4}{\partial t^4}\int_0^t \{\sin(\alpha v(t-u)) - \alpha v(t-u)\cos(\alpha v(t-u))\}\, h_c(u)\, du$$

$$= \int_0^t \left\{-3(\alpha v)^4 \sin(\alpha v(t-u)) - (\alpha v)^5 (t-u)\cos(\alpha v(t-u))\right\}$$

$$\times h_c(u)\, du + 2(\alpha v)^3 h_c(t). \qquad [1.73]$$

By combining equations [1.71]–[1.73], the following equation is obtained:

$$\left(\frac{\partial^4}{\partial t^4} + 2(\alpha v)^2 \frac{\partial^2}{\partial t^2} + (\alpha v)^4 - \lambda^4\right) h_c(t)$$

$$= 3\lambda^4 \left(\sum_{k=0}^{3} \frac{(\lambda t)^k}{k!} \right) \frac{\sin(\alpha vt) - \alpha vt \cos(\alpha vt)}{(\alpha vt)^3}.$$

The initial conditions for $h_c(t)$ are given by $h_c(0) = 0$, $\left.\frac{\partial h_c(t)}{\partial t}\right|_{t=0} = 0$, $\left.\frac{\partial^2 h_c(t)}{\partial t^2}\right|_{t=0} = 0$ and $\left.\frac{\partial^3 h_c(t)}{\partial t^3}\right|_{t=0} = 0$, which can be obtained directly from equation [1.71].

Passing to the inverse Fourier transform of equation [1.74], for $0 \leq x \leq vt$ we have

$$\left(\frac{\partial^4}{\partial t^4} - 2v^2 \sum_{i=1}^{5} \frac{\partial^4}{\partial x_i^2 \partial t^2} + v^4 \sum_{1 \leq i < j \leq 5} \frac{\partial^4}{\partial x_i^2 \partial x_j^2} - \lambda^4 \right) g(t, \mathbf{x})$$

$$= \lambda^4 \left(\sum_{k=0}^{3} \frac{(\lambda t)^k}{k!} \right) \delta_{S_5(vt)}, \qquad [1.74]$$

where $g(t, \mathbf{x}) = \mathcal{F}^{-1}(h_c(t), \alpha)$.

After factorizing equation [1.74], we obtain:

$$\left(\frac{\partial^2}{\partial t^2} - v^2 \sum_{i=1}^{5} \frac{\partial^2}{\partial x_i^2} + \lambda^2 \right) \left(\frac{\partial^2}{\partial t^2} - v^2 \sum_{i=1}^{5} \frac{\partial^2}{\partial x_i^2} - \lambda^2 \right) g(t, \mathbf{x})$$

$$= \lambda^4 \left(\sum_{k=0}^{3} \frac{(\lambda t)^k}{k!} \right) \delta_{S_5(vt)}. \qquad [1.75]$$

Since $e^{\lambda t} c_5(t, \mathbf{x}) = g(t, \mathbf{x})$, then equation [1.70] follows from equation [1.75].

The left-hand factor of equation [1.70] is the five-dimensional version of the Heaviside telegraph equation and the other factor is the five-dimensional version of the Goldstein–Kac telegraph equation.

Under the Kac condition, i.e. when $\frac{v^2}{\lambda} \to \sigma^2$, $\lambda \to +\infty$, and dividing the equation [1.70] by λ^3, we obtain the five-dimensional diffusion equation:

$$\frac{\partial}{\partial t} w(t, \mathbf{x}) - \frac{\sigma^2}{2} \sum_{i=1}^{5} \frac{\partial^2}{\partial x_i^2} w(t, \mathbf{x}) = 0.$$

1.5. The jump telegraph process in $\mathbb{R}^n$

Let us consider a similar mathematical framework as before, i.e. let the following renewal process $\xi(t) = \max\{m \geq 0 : \tau_m \leq t\}$, $t \geq 0$, where $\tau_m = \sum_{k=0}^{m} \theta_k$, $\tau_0 = 0$ and $\theta_k \geq 0$, $k = 1, 2, \ldots$, are i.i.d. random variables with cdf $G(t)$ and corresponding pdf $g(t) = \frac{d}{dt}G(t)$. Suppose that a particle starting from $(0, 0, \ldots, 0) \in \mathbb{R}^n$, at $t = 0$, continues its motion with an absolute velocity v along a direction $\boldsymbol{\eta}_0^{(n)}$, where $\boldsymbol{\eta}_0^{(n)} = (x_1, x_2, \cdots, x_n)$ is a random vector in $\mathbb{R}^n$, uniformly distributed on the unit sphere

$$\Omega_1^{n-1} = \left\{(x_1, x_2, \cdots, x_n) : x_1^2 + x_2^2 + \ldots + x_n^2 = 1\right\}.$$

At instant τ_1, the particle changes its direction to $\boldsymbol{\eta}_1^{(n)}$ and jumps by a random vector $\boldsymbol{\beta}_1$, where $\boldsymbol{\eta}_1^{(n)}$, $\boldsymbol{\eta}_0^{(n)}$ and $\boldsymbol{\beta}_1 \in \mathbb{R}^n$ are independent. Then, at time τ_2 the particle changes its direction to $\boldsymbol{\eta}_2^{(n)}$ and jumps by $\boldsymbol{\beta}_2$, where $\boldsymbol{\eta}_2^{(n)}, \boldsymbol{\eta}_1^{(n)}, \boldsymbol{\eta}_0^{(n)}, \boldsymbol{\beta}_1$ and $\boldsymbol{\beta}_2$ are independent, and so on. We assume that all $\boldsymbol{\eta}_i^{(n)}$, $i = 0, 1, 2, \ldots$ are identically distributed and all $\boldsymbol{\beta}_i$, $i = 1, 2, \ldots$ are identically distributed and radial (isotropic).

Denoting by $\mathbf{x}^{(n)}(t)$, $t \geq 0$, the particle position at time t, we then have

$$\mathbf{x}^{(n)}(t) = \sum_{j=1}^{\xi(t)} \left\{v\boldsymbol{\eta}_{j-1}^{(n)} (\tau_j - \tau_{j-1}) + \boldsymbol{\beta}_j\right\} + v\boldsymbol{\eta}_{\xi(t)}^{(n)} \left(t - \tau_{\xi(t)}\right). \quad [1.76]$$

Here, we have assumed $\sum_{j=1}^{0} = 0$.

Consider $\varphi(t, \alpha) = \mathbf{E}\left[e^{itv\left(\boldsymbol{\alpha}, \boldsymbol{\eta}_0^{(n)}\right)}\right]$, where $\alpha = \|\boldsymbol{\alpha}\|$. We should note that the function $\varphi(t, \alpha)$ is well known (Pogorui and Rodríguez-Dagnino 2012) and is of the following form:

$$\varphi(t, \alpha) = 2^{\frac{n-2}{2}} \Gamma\left(\frac{n}{2}\right) \frac{J_{\frac{n-2}{2}}(\alpha t v)}{(\alpha t v)^{\frac{n-2}{2}}}. \quad [1.77]$$

THEOREM 1.8.– The characteristic function $H(t, \alpha) = \mathbf{E}\left[e^{itv\left(\boldsymbol{\alpha}, \mathbf{x}^{(n)}(t)\right)}\right]$, $t \geq 0$, is a solution of the following integral Volterra equation of a convolution type:

$$H(t, \alpha) = (1 - G(t))\, \varphi(t, \alpha)$$

$$+\varphi_\beta(\alpha) \int_0^t g(u)\, \varphi(u, \alpha)\, H(t - u, \alpha)\, du. \quad [1.78]$$

PROOF.– It follows from equation [1.76] that:

$$H(t,\alpha) = \mathbf{E}\Big[\exp\Big\{i\Big(\boldsymbol{\alpha},\mathbf{x}^{(n)}(t)\Big)\Big\}\Big] = \mathbf{E}\Big[\exp\Big\{i\Big(\boldsymbol{\alpha},\mathbf{x}^{(n)}(t)\Big)\Big\}\mathbb{I}\{\theta_1>t\}\Big]$$

$$+\mathbf{E}\Big[\exp\Big\{i\Big(\boldsymbol{\alpha},\mathbf{x}^{(n)}(t)\Big)\Big\}\mathbb{I}\{\theta_1\le t\}\Big] = \mathbf{E}\Big[\exp\Big\{i\Big(\boldsymbol{\alpha},v\boldsymbol{\eta}_0^{(n)}t\Big)\Big\}\mathbb{I}\{\theta_1>t\}\Big]$$

$$+\mathbf{E}\Big[\exp\Big\{i\Big(\boldsymbol{\alpha},v\boldsymbol{\eta}_0^{(n)}\theta_1+\boldsymbol{\beta}_1+S_\xi+v\boldsymbol{\eta}_{\xi(t)}^{(n)}\left(t-\tau_{\xi(t)}\right)\Big)\Big\}\mathbb{I}\{\theta_1\le t\}\Big]$$

$$=(1-G(t))\,\mathbf{E}\Big[e^{itv\left(\boldsymbol{\alpha},\boldsymbol{\eta}_0^{(n)}\right)}\Big]+\int_0^t g(u)\,\mathbf{E}\Big[e^{iuv\left(\boldsymbol{\alpha},\boldsymbol{\eta}_0^{(n)}\right)}\Big]$$

$$\times\varphi_\beta(\alpha)H(t-u,\alpha)\,du,$$

where $S_\xi=\sum_{j=2}^{\xi(t)}\left(v\boldsymbol{\eta}_j^{(n)}\left(\tau_{j+1}-\tau_j\right)+\boldsymbol{\beta}_j\right)$, and $\varphi_\beta(\alpha)=\mathbf{E}\left[e^{i(\boldsymbol{\alpha},\boldsymbol{\beta}_1)}\right]$.

Denote by $f_n(t,\mathbf{x})$ the pdf of the particle position at time t. By using the n-dimension inverse Fourier transform $\mathcal{F}^{-1}$ with respect to α, we get $f_n(t,\mathbf{x})=\mathcal{F}^{-1}(H(t,\alpha))$.

1.5.1. *The jump telegraph process in* $\mathbb{R}^3$

Let us consider the three-dimensional case, that is $n=3$ and $\varphi(t,\alpha)=\frac{\sin(\alpha tv)}{\alpha tv}$. For the Erlang-2 distribution $g(t)=\lambda^2te^{-\lambda t}\mathbb{I}_{\{t\ge 0\}}$, we have

$$H(t,\alpha)=\left(e^{-\lambda t}+\lambda te^{-\lambda t}\right)\frac{\sin(\alpha tv)}{\alpha tv}$$

$$+\frac{\lambda^2\varphi_\beta(\alpha)}{\alpha v}\int_0^t e^{-\lambda u}\sin(\alpha uv)H(t-u,\alpha)\,du.$$

By denoting $h(t,\alpha)=H(t,\alpha)\,e^{\lambda t}$, we obtain

$$h(t,\alpha)=(1+\lambda t)\frac{\sin(\alpha tv)}{\alpha tv}$$

$$+\frac{\lambda^2\varphi_\beta(\alpha)}{\alpha v}\int_0^t \sin(\alpha(t-u)v)h(u,\alpha)\,du. \qquad [1.79]$$

By differentiating equation [1.79] twice, we obtain

$$\frac{\partial^2}{\partial t^2}h(t,\alpha)+\left(\alpha^2v^2-\lambda^2\varphi_\beta(\alpha)\right)h(t,\alpha)$$

$$=2\frac{\sin(\alpha tv)-\alpha tv\cos(\alpha tv)}{\alpha t^3v}, \qquad [1.80]$$

with initial conditions $h(0,\alpha) = 1$, $\frac{\partial}{\partial t} h(0,\alpha) = \lambda$, which can be obtained directly from equation [1.79].

After solving equation [1.80] with the initial conditions, we get

$$H(t,\alpha) = e^{-\lambda t} h(t,\alpha) = \frac{\sin(\alpha t v)}{\alpha t v} e^{-\lambda t} + \frac{\sin(At)}{A} \lambda e^{-\lambda t}$$

$$+ \left[\ln \frac{\alpha v - A}{\alpha v + A} + (\mathrm{Si}(t(A - \alpha v)) - \mathrm{Si}(t(A + \alpha v))) \sin(At) \right.$$

$$\left. + (\mathrm{Ci}(t(A - \alpha v)) - \mathrm{Ci}(t(A + \alpha v))) \cos(At) \right] \frac{\lambda^2 \varphi_\beta(\alpha)}{2A\alpha v} e^{-\lambda t},$$

where $A = \sqrt{\alpha^2 v^2 - \lambda^2 \varphi_\beta(\alpha)}$, $\mathrm{Si}(x) = \int_0^x \frac{\sin(t)}{t} dt$, $\mathrm{Ci}(x) = \gamma + \ln(x) + \int_0^x \frac{\cos(t)-1}{t} dt$, and γ is the Euler–Mascheroni constant.

1.5.2. *Conclusions and final remarks*

The core of our analysis is based on integral equations for the characteristic function of the movement of a particle. For the Erlang case used in this work, the proof of the integral equation was carried out in two different ways. First, in Pogorui and Rodríguez-Dagnino (2011), the proof was carried out in a detailed manner and it reduces to an exponential or Markov case, which is well accepted in other published papers. Second, in Pogorui and Rodríguez-Dagnino (2013), the proof was carried out on the basis of renewal theory and corresponds to the semi-Markov case.

Moreover, our results obtained for 2D do not contradict known results stating that a continuous part of the density is a solution of a two-dimensional telegraph equation with a zero right-hand side. However, we must remark that this statement is not the complete view of this problem. For instance, for $|x| < vt$, the right-hand side is also zero. At $|x| = vt$, the continuous part of the generalized pdf has singularity (it goes to infinity as $|x| \to vt$), and it has the so-called explosion effect for just the continuous part of the density. In addition, in our approach we used the integral equation for the characteristic function of motion on the plane for the Markov case, and we formulated the characteristic function for the motion density as the sum of the characteristic function of the singular part and the continuous part, which is barely considered in other works. Some recent results in this area are associated with the Dirichlet distribution (see De Gregorio and Orsingher 2012). However, it is not clear that by considering the Dirichlet distribution, the three-dimensional telegraph equation will appear as natural as it does for the Erlang case, and as a consequence it is not clear that our explanation of the physical process related to three-dimensional telegraph equation can be obtained from the Dirichlet distribution, as it has been done in our work. The physical relevance of our analysis can be appreciated after looking at the physical models in Tautz and Lerche (2016) and Morse and Feshbach (1953). We have to mention that the incomplete distribution density for the position of such a process described by the three-dimensional telegraph equation, was recently obtained in Tautz and Lerche (2016).

Furthermore, we are including plots of the main results to clarify some of the theoretical issues stated in this work. We are also including the five-dimensional case in our analysis, which may have some theoretical interest.

2

System of Interactive Particles with Markov and Semi-Markov Switching

In this chapter, we study the system of telegraph processes with interaction, which can be interpreted as a model of a one-dimensional gas. In some ways, this system is a generalization of dynamical systems of billiard type (Kornfeld *et al.* 1982). We have developed the distribution of the meeting instant of two telegrapher particles on the line (or collision phenomenon) that started at the same time from different positions on the line. The distribution of particle positions in a fixed time for systems with reflecting boundaries and without boundaries, and limiting properties of this distribution, are studied. In this model, we also study the cumulative time of the free path for the system of particles with an elastic collision, and this result extends those obtained for the family of Wiener processes (Dorogovtsev 2007a) to the case of the telegraph processes, i.e. to a model of a one-dimensional gas, whose molecules have a finite speed.

Moreover, in section 2.1, we construct a system of telegraph processes with interactions. This random motion can be interpreted as a model of an ideal gas, which is a kind of generalization of the one-dimensional deterministic models of gas studied, for example, by Kornfeld *et al.* (1982), Sinai (2010), Vysotsky (2008b,a), and others.

We also present further results of this model, i.e. we investigate the free path times of a family of particles before they experience collision with any other particle. We also study the distribution of particles, which are described by telegraph processes with elastic collisions and reflecting boundaries, and its limiting properties.

In section 2.1.1, we systematically present the distribution of the meeting instant of two telegraph particles (processes) on the line that started their motion at the same time but from different locations. We study the limit of this distribution under Kac's condition.

In section 2.1.4, we develop the equations to model a system of particles with a Markov switching process and boundaries with rejection. We obtain distributions of particle positions and study the limiting properties of these distributions.

In section 2.1.5, we obtain the estimation of the number of collisions in a system of interacting particles without boundaries.

In section 2.2, many of these results of the interaction of particles have been extended to semi-Markov processes.

2.1. Description of the Markov model

Let $\{\xi(t), t \geq 0\}$ be a Markov process on the phase space $\{0, 1\}$ with generating matrix

$$Q = \lambda \begin{pmatrix} -1 & 1 \\ 1 & -1 \end{pmatrix}.$$

It is said that $S(t)$ is a telegraph process if

$$\frac{d}{dt} S(t) = v(-1)^{\xi(t)}, \quad v = const > 0, \qquad S(0) = y_0.$$

1) For a set of ordered real numbers $y_1 < y_2 < \cdots < y_n$ consider a family of independent telegraph processes $S_i(t)$, $i = 1, 2, \ldots, n$, $t \geq 0$, $S_i(0) = y_i$, with the same absolute velocity v and a parameter of switching process λ.

2) Let the process $x(y_i, t)$, $t \geq 0$, describing the position of the particle i at time t, be in coincidence with the order statistics $S_{(i)}(t)$ of the processes $S_i(t)$, $i = 1, 2, \ldots, n$, that is

$$x(y_i, t) = S_{(i)}(t).$$

This approach for the specification of particle motion means that each particle of the system is moving in accordance with the respective telegraph process, before an elastic collision with another particle. For an elastic collision of two particles, we understand a collision in which the particles change direction to the opposite, i.e. exchanging momentum. In terms of the processes, this means that the particles exchange the telegraph processes that determined their movement.

REMARK 2.1.– *We should note that the processes $x(y_i, t)$, $i = 1, 2, \ldots, n$ are not telegraph for each $t \geq 0$.*

REMARK 2.2.– *It follows from the description of $x(y,t)$ that $x(y_1,t) \leq x(y_2,t) \leq \cdots \leq x(y_n,t)$ for any $t \in [0,T]$.*

Models for Wiener processes with several types of coalescence after collision are called Harris flows, and such processes have been of interest in many research works (e.g. (Dorogovtsev 2007b; Arratia 1979; Dorogovtsev 2007a, 2010; Konarovskii 2011; Le and Raimond 2004; Pilipenko 1999)).

For these models, we consider the following functionals and their distribution: the number of collisions in a system of interacting particles both with reflecting boundaries and without them; joint time of free paths for the family of particles with opposite initial velocities and elastic collision; the distribution of particle positions of the model in a fixed time and their limit properties. It should be noted that, as we know from physics, the length of the free path of gas molecules is inversely proportional to the pressure at a fixed temperature, and the number of collisions of gas molecules is proportional to the average speed of the molecules, which determines the gas temperature or energy. Therefore, functionals of a gas model, which are studied in this section, are important in the study of the physical properties of gas.

2.1.1. *Distribution of the first meeting time of two telegraph processes*

In this section, we have developed the distribution of the first meeting instant of two telegraph processes (or the trajectory of the motion of two particles modeled by telegraph processes) on the line (or collision phenomenon) that started at the same time from different positions on the line.

Consider two particles, 1 and 2, on a line. Suppose $S_i(t) = y_i + \int_0^t (-1)^{\xi_i(s)} ds$, $i = 1, 2$, where $\{\xi_i(t),\, t \geq 0\}$ are two independent Markov processes on the phase space $\{0, 1\}$, and with infinitesimal matrix

$$Q = \lambda \begin{pmatrix} -1 & 1 \\ 1 & -1 \end{pmatrix}.$$

We also have $x(y_i, t) = S_{(i)}(t)$, $i = 1, 2$.

We assume that $z = y_2 - y_1 > 0$ and denote by $\Delta(t) = x(y_2, t) - x(y_1, t)$.

Let us introduce the bivariate process $\eta(t) = (\xi_1(t), \xi_2(t))$, and consider the following random variable:

$$\tau_{(k_1,k_2)}(z) = \inf\{t \geq 0 : \Delta(t) = 0,\ \eta(0) = (k_1, k_2)\},\ \ k_j \in \{0, 1\}.$$

It is easily seen that $\tau_{(k_1,k_2)}(z)$ is the instant of the first particles collision assuming $\eta(0) = (k_1, k_2)$.

Denote by $f_{(k_1,k_2)}(t,z)\,dt = P\left(\tau_{(k_1,k_2)}(z) \in dt\right)$ the pdf of $\tau_{(k_1,k_2)}(z)$. Our purpose is to calculate the pdf of the random variable $\tau_{(k_1,k_2)}(z)$.

THEOREM 2.1.– For $t \geq \frac{z}{2v}$, we have

$$f_{(0,1)}(t,z) = e^{-2\lambda t}\delta(z - 2vt) + \frac{z\lambda}{2v^2}e^{-2\lambda t}\frac{I_1\left(\frac{\lambda}{v}\sqrt{4v^2t^2 - z^2}\right)}{\sqrt{4v^2t^2 - z^2}}, \qquad [2.1]$$

$$f_{(0,0)}(t,z) = f_{(1,1)}(t,z)$$
$$= \frac{z\lambda}{2v^2}e^{-2\lambda t}\int_{z/2v}^{t}\frac{I_1\left(\frac{\lambda}{v}\sqrt{4v^2u^2 - z^2}\right)}{\sqrt{4v^2u^2 - z^2}}\frac{I_1(2\lambda(t-u))}{(t-u)}du \qquad [2.2]$$

and

$$f_{(1,0)}(t,z) = \frac{z\lambda}{2v^2}e^{-2\lambda t}\int_{z/2v}^{t}\frac{I_1\left(\frac{\lambda}{v}\sqrt{4v^2u^2 - z^2}\right)}{\sqrt{4v^2u^2 - z^2}}$$
$$\times\int_0^{t-u}\frac{I_1(2\lambda(t-u-v))}{(t-u-v)}\frac{I_1(2\lambda v)}{v}dvdu. \qquad [2.3]$$

PROOF.– Let us consider the Laplace transforms of $\tau_{(k_1,k_2)}(z)$, $k_i \in \{0,1\}$.

$$\varphi_{(k_1,k_2)}(s,z) = \mathbf{E}\left[e^{-s\tau_{(k_1,k_2)}(z)}\right], \quad s > 0.$$

By using a fundamental method of renewal theory, we can write the following system of integral equations for these Laplace transforms:

$$\varphi_{(0,1)}(s,z) = e^{-\frac{s+2\lambda}{2v}z} + \frac{\lambda}{2v}\int_0^z e^{-\frac{s+2\lambda}{2v}u}\varphi_{(1,1)}(s,z-u)\,du$$
$$+\frac{\lambda}{2v}\int_0^z e^{-\frac{s+2\lambda}{2v}u}\varphi_{(0,0)}(s,z-u)\,du$$
$$= e^{-\frac{s+2\lambda}{2v}z} + \frac{\lambda}{2v}e^{-\frac{s+2\lambda}{2v}z}\int_0^z e^{\frac{s+2\lambda}{2v}u}\left(\varphi_{(1,1)}(s,u) + \varphi_{(0,0)}(s,u)\right)du,$$

$$\varphi_{(0,0)}(s,z) = \lambda\int_0^\infty e^{-(s+2\lambda)u}\varphi_{(0,1)}(s,z)\,du + \lambda\int_0^\infty e^{-(s+2\lambda)u}\varphi_{(1,0)}(s,z)\,du$$
$$= \frac{\lambda}{s+2\lambda}\left(\varphi_{(0,1)}(s,z) + \varphi_{(1,0)}(s,z)\right),$$

$$\varphi_{(1,1)}(s,z) = \lambda \int_0^\infty e^{-(s+2\lambda)u} \varphi_{(0,1)}(s,z)\, du + \lambda \int_0^\infty e^{-(s+2\lambda)u} \varphi_{(1,0)}(s,z)\, du$$

$$= \frac{\lambda}{s+2\lambda}\left(\varphi_{(0,1)}(s,z) + \varphi_{(1,0)}(s,z)\right),$$

$$\varphi_{(1,0)}(s,z) = \frac{\lambda}{2v}\int_0^\infty e^{-\frac{s+2\lambda}{2v}u}\left(\varphi_{(0,0)}(s,z+u) + \varphi_{(1,1)}(s,z+u)\right) du$$

$$= \frac{\lambda}{2v} e^{\frac{s+2\lambda}{2v}z}\int_z^\infty e^{-\frac{s+2\lambda}{2v}u}\left(\varphi_{(0,0)}(s,u) + \varphi_{(1,1)}(s,u)\right) du.$$

It is easily seen that

$$\varphi_{(0,0)}(s,z) = \varphi_{(1,1)}(s,z). \qquad [2.4]$$

Taking into account that

$$\varphi_{(0,0)}(s,z) + \varphi_{(1,1)}(s,z) = \frac{2\lambda}{s+2\lambda}\left(\varphi_{(0,1)}(s,z) + \varphi_{(1,0)}(s,z)\right), \qquad [2.5]$$

we have

$$\frac{\partial}{\partial z}\varphi_{(0,1)}(s,z) = -\frac{s+2\lambda}{2v}\varphi_{(0,1)}(s,z)$$

$$+\frac{\lambda^2}{v(s+2\lambda)}\left(\varphi_{(0,1)}(s,z) + \varphi_{(1,0)}(s,z)\right),$$

$$\frac{\partial}{\partial z}\varphi_{(1,0)}(s,z) = \frac{s+2\lambda}{2v}\varphi_{(1,0)}(s,z)$$

$$-\frac{\lambda^2}{v(s+2\lambda)}\left(\varphi_{(0,1)}(s,z) + \varphi_{(1,0)}(s,z)\right).$$

It is well known that $\varphi_{(1,0)}(s,z)$ and $\varphi_{(0,1)}(s,z)$ satisfy the following equation:

$$\det\begin{pmatrix} \frac{\partial}{\partial z} + \frac{(s+2\lambda)}{2v} - \frac{\lambda^2}{v(s+2\lambda)} & -\frac{\lambda^2}{v(s+2\lambda)} \\ \frac{\lambda^2}{v(s+2\lambda)} & \frac{\partial}{\partial z} - \frac{(s+2\lambda)}{2v} + \frac{\lambda^2}{v(s+2\lambda)} \end{pmatrix} f(z) = 0.$$

By calculating the determinant, we have

$$\frac{\partial^2}{\partial z^2} f(z) - \frac{s^2+4\lambda s}{4v^2} f(z) = 0.$$

Solving this equation, we have

$$f(z) = C_1 e^{\sqrt{s^2+4\lambda s}\,\frac{z}{2v}} + C_2 e^{-\sqrt{s^2+4\lambda s}\,\frac{z}{2v}}.$$

The constants C_1 and C_2 obtained from the system of integral equations yields

$$\varphi_{(0,1)}(s,z) = e^{-\frac{z}{2v}\sqrt{s^2+4\lambda s}}, \qquad [2.6]$$

and

$$\varphi_{(1,0)}(s,z) = \frac{s+2\lambda-\sqrt{s^2+4\lambda s}}{s+2\lambda+\sqrt{s^2+4\lambda s}} e^{-\frac{z}{2v}\sqrt{s^2+4\lambda s}}. \qquad [2.7]$$

Taking into account equations [2.4] and [2.5], we have

$$\varphi_{(0,0)}(s,z) = \varphi_{(1,1)}(s,z) = \frac{s+2\lambda-\sqrt{s^2+4\lambda s}}{2\lambda} e^{-\frac{z}{2v}\sqrt{s^2+4\lambda s}}. \qquad [2.8]$$

The inverse Laplace of $\varphi_{(0,1)}(s,z)$ with respect to s yields the pdf (Korn and Korn 2000, no. 88, p. 239),

$$f_{(0,1)}(t,z) = \mathcal{L}^{-1}\left(e^{-\frac{z}{2v}\sqrt{s^2+4\lambda s}}, t\right)$$

$$= e^{-2\lambda t}\delta(z-2vt) + 2z\lambda e^{-2\lambda t}\frac{I_1\left(\frac{\lambda}{v}\sqrt{4v^2t^2-z^2}\right)}{\sqrt{4v^2t^2-z^2}}, \quad t \geq \frac{z}{2v}. \qquad [2.9]$$

Hence, for $t \geq \frac{z}{2v}$

$$P\left(\tau_{(0,1)}(z) \in dt\right) = e^{-2\lambda t}\delta(z-2vt)\,dt + 2z\lambda e^{-2\lambda t}\frac{I_1\left(\frac{\lambda}{v}\sqrt{4v^2t^2-z^2}\right)}{\sqrt{4v^2t^2-z^2}}dt.$$

It is easily verified that

$$e^{-\frac{z}{2v}\sqrt{s^2+4\lambda s}} = e^{z\left[-\frac{s}{2v}+\int_0^\infty \left(1-e^{-sy}\right)\frac{\lambda}{v}\frac{e^{-2\lambda y}I_1(2\lambda y)}{y}dy\right]}. \qquad [2.10]$$

Then, from Bateman (1954, no. 49, p. 237), we obtain the following inverse Laplace transform:

$$\mathcal{L}^{-1}\left(1+\frac{s-\sqrt{s^2+4\lambda s}}{2\lambda}, t\right) = e^{-2\lambda t}\frac{I_1(2\lambda t)}{t}. \qquad [2.11]$$

Hence, it is easily seen that the following condition holds:

$$\int_0^\infty \min(1,t)\, e^{-2\lambda t}\frac{I_1(2\lambda t)}{t}dt < +\infty. \qquad [2.12]$$

It is well known (Meyer 1969) that the distribution, in which the Laplace transform can be represented as the right side of equation [2.10] with the condition [2.12], belongs to the infinitely divisible distribution.

Therefore, the pdf $f_{(0,1)}(t,z)$ is the infinitely divisible density function.

By using formula 49 of Bateman (1954, p. 237), we have

$$\mathcal{L}^{-1}\left(\frac{s+2\lambda-\sqrt{s^2+4\lambda s}}{s+2\lambda+\sqrt{s^2+4\lambda s}},t\right)=\frac{1}{4\lambda^2}\mathcal{L}^{-1}\left(\left(s+2\lambda-\sqrt{s^2+4\lambda s}\right)^2,t\right)$$

$$=e^{-2\lambda t}\int_0^t\frac{I_1(2\lambda(t-v))}{(t-v)}\frac{I_1(2\lambda v)}{v}dv.$$

By calculating $\mathcal{L}^{-1}\left(\frac{s+2\lambda-\sqrt{s^2+4\lambda s}}{s+2\lambda+\sqrt{s^2+4\lambda s}}e^{-\frac{z}{2v}\sqrt{s^2+4\lambda s}}\right)$, we obtain equation [2.3].

It is easily seen that $f_{(0,1)}(t,z)$ is a heavy tail probability density function with respect to t. Indeed, by using an asymptotic expansion for $I_1(t)$ (Gradshteyn and Ryzhik 1980), we have

$$\lim_{t\to+\infty}\sqrt{2\pi t}I_1(t)e^{-t}=1, \qquad [2.13]$$

and therefore, when $\alpha\geq\frac{1}{2}$

$$\mathbf{E}\left[\{\tau_{(0,1)}(z)\}^{\alpha}\right]\geq 2z\lambda\int_{\frac{z}{2v}}^{\infty}t^{\alpha}e^{-2\lambda t}\frac{I_1\left(\frac{\lambda}{v}\sqrt{4v^2t^2-z^2}\right)}{\sqrt{4v^2t^2-z^2}}dt=+\infty.$$

Considering [2.13], it is easily verified that $\mathbf{E}\left[\{\tau_{(0,1)}(z)\}^{\alpha}\right]<\infty$ for $0\leq\alpha<\frac{1}{2}$.

For $\tau_{(0,1)}(z)$ at time $t=0$, the particles move in opposite directions to meet each other and for $\tau_{(1,0)}(z)$ at time $t=0$, the particles move in opposite directions far away from each other.

Hence, $\mathbf{E}\left[\tau_{(0,1)}(z)\right]\leq\mathbf{E}\left[\tau_{(1,0)}(z)\right]$ and $f_{(1,0)}(t,z)$ is also a heavy tail density function with respect to t.

Let us consider the following so-called Kac's condition (or the hydrodynamic limit): denote by $\lambda=\varepsilon^{-2}$, $v=c\varepsilon^{-1}$, as $\varepsilon>0$, that is $v>+\infty$, and $\lambda>+\infty$, such that $\frac{v^2}{\lambda}>c^2$.

It was proved (Kac 1974) that under Kac's condition the telegraph process $x(t)$ weakly converges to the Wiener process $\{w(t)\}$.

Denote by $f(t,z)=\dfrac{cz\exp\left(-\frac{c^2z^2}{4t}\right)}{2\sqrt{\pi}t^{3/2}}$. It is well known that $f(t,z)$ is the pdf of a collision instant of two particles moving according to Wiener paths $w(t)$, where $z>0$ is the distance between starting points of the particles.

LEMMA 2.1.– For each $k_1, k_2 \in \{0,1\}$, the distribution function with pdf $f_{(k_1, k_2)}(t, z)$ weakly converges to the distribution function with pdf $f(t, z)$.

PROOF.– It follows from equations [2.6] to [2.8] that

$$\lim_{\varepsilon \to 0} \varphi_{(k_1, k_2)}(s, z) = e^{-zc\sqrt{s}}.$$

Applying the inverse Laplace transform, we have

$$\mathcal{L}^{-1}\left(e^{-zc\sqrt{s}}\right) = \frac{cz \exp\left(-\frac{c^2 z^2}{4t}\right)}{2\sqrt{\pi} t^{3/2}} = f(t, z).$$

Therefore, under Kac's conditions not only the telegraph process weakly converges to the Wiener process, but also the first meeting instant of two telegraph processes weakly converges to the first meeting instant of the corresponding two Wiener processes.

2.1.2. *Estimate of the number of particle collisions*

Denote by $C_{(0,1)}(t, z)$ the number of collisions of particles $x(y_i, t), i = 1, 2$ during time $(0, t)$, $t > 0$ provided that $\eta(0) = (0, 1)$. It is easily verified that $C_{(0,1)}(t, z)$ is the number of intersections $I_{(0,1)}(t, y_j - y_i)$ of the telegraph processes $S_i(t)$ and $S_j(t)$ provided that $(\xi_i(0), \xi_j(0)) = (0, 1)$.

Consider the renewal function $H_{(0,1)}(t, z) = \mathbf{E}[C_{(0,1)}(t, z)]$. By using the Laplace transform for general renewal function (Cox 1970), it follows from equations [2.6] and [2.7] that the Laplace transform $\hat{H}_{(0,1)}(s, z) = \mathcal{L}\left(H_{(0,1)}(t, z), t\right)$ of $H_{(0,1)}(t, z)$ with respect to t has the following form:

$$\hat{H}_{(0,1)}(s, z) = \frac{e^{-\frac{z}{2v}\sqrt{s^2+4\lambda s}}}{s} \sum_{k=0}^{\infty} \left(\frac{s + 2\lambda - \sqrt{s^2 + 4\lambda s}}{s + 2\lambda + \sqrt{s^2 + 4\lambda s}}\right)^k$$

$$= e^{-\frac{z}{2v}\sqrt{s^2+4\lambda s}} \left(\frac{s + 2\lambda + \sqrt{s^2 + 4\lambda s}}{2s\sqrt{s^2 + 4\lambda s}}\right).$$

It is easily verified that

$$\mathcal{L}^{-1}\left(\frac{s + 2\lambda + \sqrt{s^2 + 4\lambda s}}{2s\sqrt{s^2 + 4\lambda s}}\right)$$

$$= \frac{1}{2} + \left(\left(\frac{1}{2} + \lambda t\right) I_0(2\lambda t) + \lambda t I_1(2\lambda t)\right) e^{-2\lambda t}.$$

Therefore,

$$H_{(0,1)}(t,z)=\int_{\frac{z}{2v}}^{t}e^{-2\lambda u}\left(\delta(z-2vu)+\frac{z\lambda}{2v^2}\frac{I_1\left(\frac{\lambda}{v}\sqrt{4v^2u^2-z^2}\right)}{\sqrt{4v^2u^2-z^2}}\right)$$

$$\times\left(\frac{1}{2}-e^{-2\lambda(t-u)}\left(\left(\frac{1}{2}+\lambda(t-u)\right)I_0(2\lambda(t-u))\right.\right.$$

$$\left.\left.+\lambda(t-u)I_1(2\lambda(t-u))\right)\right)du$$

$$=\frac{e^{-\frac{\lambda z}{v}}}{2}+\frac{z\lambda}{4v^2}\int_{\frac{z}{2v}}^{t}e^{-2\lambda u}\frac{I_1\left(\frac{\lambda}{v}\sqrt{4v^2u^2-z^2}\right)}{\sqrt{4v^2u^2-z^2}}du$$

$$+e^{-2\lambda t}\left(\left(\frac{1}{2}+\lambda\left(t-\frac{z}{2v}\right)\right)I_0\left(2\lambda\left(t-\frac{z}{2v}\right)\right)\right. \qquad [2.14]$$

$$\left.+\lambda\left(t-\frac{z}{2v}\right)I_1\left(2\lambda\left(t-\frac{z}{2v}\right)\right)\right)+e^{-2\lambda t}\frac{z\lambda}{4v^2}\int_{\frac{z}{2v}}^{t}\frac{I_1\left(\frac{\lambda}{v}\sqrt{4v^2u^2-z^2}\right)}{\sqrt{4v^2u^2-z^2}}$$

$$\times\left(1+\left((1+2\lambda(t-u))I_0(2\lambda(t-u))+\lambda(t-u)I_1(2\lambda(t-u))\right)\right)du.$$

It follows from equation [2.14] that by defining $\lambda=\varepsilon^{-2}$, $v=c\varepsilon^{-1}$, we have

$$H_{(0,1)}(t,z)=O\left(\varepsilon^{-1}\right)=O\left(\sqrt{\lambda}\right)=O(v),\ \text{as}\quad \varepsilon\to 0.$$

Similarly, we can show that for each $k_1,\ k_2\in\{0,1\}$

$$H_{(k_1,k_2)}(t,z)=O\left(\varepsilon^{-1}\right)=O\left(\sqrt{\lambda}\right)=O(v),\ \text{as}\quad \varepsilon\to 0.$$

Denote by $C_{(k_1,k_2,\ldots,k_n)}(t,\,y_1,\,y_2,\ldots,\,y_n),\ \ k_i\in\{0,1\},\,y_1<y_2<\ldots<y_n$ the number of collisions of the particles $x(y_i,t)\,,\ i=1,2,\ldots,n$ during time $(0,t)$, $t>0$. It is assumed that $\eta(0)=(\xi_1(0)\,,\xi_2(0)\,,\ldots,\xi_n(0))=(k_1,k_2,\ldots,k_n)$, where $\{\xi_i(t),\,t\geq 0,\,i=1,\ldots,n\}$ are independent Markov processes on the phase space $\{0,1\}$ with infinitesimal matrix

$$Q=\lambda\begin{pmatrix}-1&1\\1&-1\end{pmatrix}.$$

It is easily seen that

$$H_{(k_1,k_2,\ldots,k_n)}(t,\,y_1,\,y_2,\ldots,\,y_n)=\mathbf{E}\left[C_{(k_1,k_2,\ldots,k_n)}(t,\,y_1,\,y_2,\ldots,\,y_n)\right]$$

$$=\sum_{i=1}^{n-1}H_{(k_i,\,k_{i+1})}(t,\,y_{i+1}-y_i),$$

where $H_{(k_i,k_{i+1})}(t,y_{i+1}-y_i)$ can be calculated in a similar manner as in [2.14].

2.1.3. *Free path times of a family of particles*

Suppose two telegraph particles starting from points $y < y^*$ move toward each other and their movement is determined by the processes $x(y,t)$ and $x(y^*,t)$. For a fixed $T > 0$, consider the following random time:

$$\tau_{(0,1)}^T = \min\left(T, \inf\left\{t : x(y,t) - x(y^*,t) = 0\right\}\right),$$

LEMMA 2.2.– There exists $C > 0$ such that for any two points y, y^* $(y < y^*)$ that satisfy $z = y^* - y < 2vT$, we have

$$\mathbf{E}\left[\tau_{(0,1)}^T\right] \leq C\left(y^* - y\right). \qquad [2.15]$$

PROOF.– Since $T > \frac{z}{2v}$, we have

$$\mathbf{E}\left[\tau_{(0,1)}^T\right] = \int_0^T te^{-2\lambda t}\left[\delta(z - 2vt) + 2z\lambda\frac{I_1\left(\frac{\lambda}{v}\sqrt{4v^2t^2 - z^2}\right)}{\sqrt{4v^2t^2 - z^2}}I_{\left\{t \geq \frac{z}{2v}\right\}}\right]dt$$

$$+2z\lambda T\int_T^{\infty} e^{-2\lambda t}\frac{I_1\left(\frac{\lambda}{v}\sqrt{4v^2t^2 - z^2}\right)}{\sqrt{4v^2t^2 - z^2}}dt$$

$$\leq \frac{z}{2v} + 2z\lambda\int_{\frac{z}{2v}}^T t\frac{I_1\left(\frac{\lambda}{v}\sqrt{4v^2t^2 - z^2}\right)}{\sqrt{4v^2t^2 - z^2}}dt + 2z\lambda T\int_T^{\infty} e^{-2\lambda t}\frac{I_1(2\lambda t)}{vt}dt$$

$$= \frac{z}{2v} + \frac{z}{2v}\left(I_0\left(\frac{\lambda}{v}\sqrt{4T^2v^2 - z^2}\right) - 1\right) + \frac{z\lambda T}{v}\int_0^{\infty} e^{-2\lambda t}\frac{I_1(2\lambda t)}{t}dt$$

$$\leq Cz,$$

where $C = \frac{I_0(2T\lambda)}{2v} + \frac{\lambda T}{v}$.

Consider an increasing sequence of numbers $\{0 < y_1 < y_2 < \dots < y_n\} \subset [0, L]$, $0 < L < \infty$. Suppose that the set of independent telegraph processes $\{S_k(t),\ k = 1, 2, \dots n\}$ such that the process $S_k(t)$, starting from y_k, has the initial velocity $(-1)^{k+1}v$, $k = 1, 2, \dots n$. As before, the position $x(y_k, t)$ of the particle k is determined by order statistics $x(y_k, t) = S_{(k)}(t)$.

We introduce the following random times:

$$\tau_{n1} = \min\left(T, \inf\left\{t : \left(x(y_2,t) - x(y_1,t)\right) = 0\right\}\right),$$

$$\tau_{nk} = \min\left(T, \inf\left\{t : \left(x(y_k,t) - x(y_{k-1},t)\right)\left(x(y_k,t) - x(y_{k+1},t)\right) = 0\right\}\right),$$
$$2 \leq k \leq n-1,$$

$$\tau_{nn} = \min\left(T, \inf\left\{t : \left(x(y_n,t) - x(y_{n-1},t)\right) = 0\right\}\right).$$

The random time τ_{nk} is the time of the free path of particle k in the nth trial of a Bernoulli scheme until the collision with the neighboring particle, and it is equal to T if there is no collision.

THEOREM 2.2.– Suppose that for all $n \geq 1$ we have the sequence

$$\{0 < y_1 < y_2 < \cdots < y_n\} \subset [0, L], \quad 0 < L < \infty$$

and for all $1 \leq i \leq n-1$, $y_{i+1} - y_i < 2vT$. Then

$$\lim_{n\to\infty} \sum_{k=1}^{n} \tau_{nk} < \infty \ \text{(a.s.)}$$

PROOF.– Consider the random times for $1 \leq k \leq \left[\frac{n}{2}\right]$

$$\widetilde{\tau}_{n2k} = \widetilde{\tau}_{n2k-1} = \min\left(T, \inf\left\{t : \left(S_{2k}(t) - S_{2k-1}(t)\right) = 0\right\}\right).$$

When n is odd, we define $\widetilde{\tau}_{nn} = T$.

Since $\tau_{nk} \leq \widetilde{\tau}_{nk}$, it is sufficient to show that $\lim_{n\to\infty} \sum_{k=1}^{n} \widetilde{\tau}_{nk} < \infty$ (a.s.)

Let us prove that

$$\lim_{n\to\infty} \sum_{k=1}^{n} \mathbf{E}\left[\widetilde{\tau}_{nk}\right] < +\infty.$$

From lemma 2.2, it follows that there exists $C > 0$ such that for all $1 \leq k \leq \left[\frac{n}{2}\right]$.

$$\mathbf{E}\left[\widetilde{\tau}_{n2k}\right] = \mathbf{E}\left[\widetilde{\tau}_{n2k-1}\right] \leq C\left(y_{2k} - y_{2k-1}\right).$$

Thus,

$$\lim_{n\to\infty} \sum_{k=1}^{n} \mathbf{E}\left[\tau_{nk}\right] \leq \lim_{n\to\infty} \sum_{k=1}^{n} \mathbf{E}\left[\widetilde{\tau}_{nk}\right]$$

$$\leq 2C \lim_{n\to\infty} \sum_{i=1}^{[n/2]} \left(y_{2i} - y_{2i-1}\right) + T \leq (2CS + T).$$

From $\tau_{nk} \geq 0$, it follows that $\mathbf{E}\left[\lim_{n\to\infty} \sum_{k=1}^{n} \tau_{nk}\right] = \lim_{n\to\infty} \sum_{k=2}^{n} \mathbf{E}\left[\tau_{nk}\right]$.

Therefore,

$$\lim_{n\to\infty} \sum_{k=1}^{n} \tau_{nk} < \infty \quad \text{(a.s.)}.$$

We should note that theorem 2.2 was proved for the case of Wiener particles in Dorogovtsev (2007a) without any restriction on the initial velocities of the particles.

2.1.4. *Estimation of the number of particle collisions for systems with boundaries*

In general, for an ideal gas the number of particle collisions is directly related to the energy of the system, and it is therefore important to estimate the number of particle collisions in the case of such a gas constrained to some boundaries.

Consider a set of ordered real numbers $\{y_i;\ i = 1, \dots, n\} \subset (0,\ b)$, where $b > 0$ and $y_1 < y_2 < \dots < y_n$. Let $S_1(t), S_2(t), \dots, S_n(t)$ be a family of independent telegraph processes. We assume that all the processes have absolute velocity v and a parameter of switching process λ, and starting from y_i the process $S_i(t)$ has equal probabilities of initial directions of motion. We suppose that 0 and b are two reflecting boundaries such that if a process reaches boundary 0 or b, then it changes velocity to the opposite direction.

Consider the family of particles with trajectories

$$x(y_1, t), x(y_2, t), \dots, x(y_n, t),$$

where every $x(y_i, t)$ particle coincides, respectively, with the process $S_i(t)$ before its first elastic collision with another particle or, equivalently, to the first intersection of the process $S_i(t)$ with another process. After the first elastic collision of the particle $x(y_i, t)$ with another particle, say $x(y_j, t)$, they will switch the telegraph processes that describe their trajectories so $S_i(t)$ will coincide with the trajectory of $x(y_j, t)$, and so on. It is easily seen that the trajectories of the particles $x(y_i, t)$, $i = 1, 2, \dots$ coincide with the order statistics of $S_i(t)$, $i = 1, 2, \dots$ as follows:

$$x(y_1, t) = S_{(1)}(t),\ \ x(y_2, t) = S_{(2)}(t),\ \ \dots,\ \ x(y_n, t) = S_{(n)}(t). \qquad [2.16]$$

Let us introduce the following distribution functions:

$$F_{y_k}(x) = P\{x(y_k, t) \le x\}, \quad k = 1, 2, \dots.$$

Denote by $M_k^{(l)}$, $l = 1, 2, \dots, C_n^k$, different k-subsets of the set $M = \{1, 2, \dots, n\}$.

From [2.16], it follows that

$$F_{y_r}(x) = \sum_{k=r}^{n} \sum_{l=1}^{C_n^k} \prod_{i \in M_k^{(l)}} P(S_i(t) \le x) \prod_{j \in M \setminus M_k^{(l)}} P(S_j(t) > x).$$

For some particular cases, we have

$$F_{y_1}(x) = P(x(y_1, t) \le x) = 1 - \prod_{i=1}^{n} P(S_i(t) > x),$$

$$\vdots$$

$$F_{y_{n-1}}(x) = \sum_{k=1}^{n} \prod_{\substack{i=1 \\ i \neq k}}^{n} P(S_i(t) \leq x) P(S_k(t) > x) + \prod_{i=1}^{n} P(S_i(t) \leq x),$$

$$F_{y_n}(x) = P(x(y_n, t) \leq x) = \prod_{i=1}^{n} P(S_i(t) \leq x).$$

Let us study the limiting distribution of $S_k(t)$, $k = 1, \ldots, n$ as $t \to +\infty$.

Denote by $N(t)$ the number of Poisson events that have occurred in the interval $(0, t)$ and the s_j, $j \geq 0$ represent the instants at which Poisson events occur, and $s_0 = 0$. We assume that instants s_j denote times of change of direction of $S_k(t)$.

THEOREM 2.3.– Suppose that $f(x)$ is an integrable function on $[0, b]$. Then

$$P\left(\lim_{T \to +\infty} \frac{1}{T} \int_0^T f(S_1(t))\, dt = \frac{1}{b} \int_0^b f(x)\, dx\right) = 1.$$

PROOF.– We will use the well-known strong law of large numbers for the Poisson process

$$P\left(\lim_{T \to +\infty} \frac{N(T)}{T} = \lambda\right) = 1. \qquad [2.17]$$

Then, with probability one we have the following equations:

$$\lim_{T \to +\infty} \frac{1}{T} \int_0^T f(S_1(t))\, dt = \lim_{T \to +\infty} \frac{1}{T} \left(\sum_{i=0}^{N(T)} \int_{s_j}^{s_{j+1}} f(S_1(t))\, dt \right)$$

$$= \lim_{T \to +\infty} \frac{1}{T} \sum_{i=0}^{N(T)} \left(\left[\frac{(s_{j+1} - s_j) v}{2b} \right] \frac{2}{v} \int_0^b f(x)\, dx + r_i \right), \qquad [2.18]$$

where $r_i = \int_{u_i}^{u_i + v_i} f(S_k(t))\, dt$, and u_i, v_i are independent random variables and u_i is uniformly distributed on $[0, 2b]$, v_i has the following pdf:

$$g(t) = \frac{\lambda}{v} e^{-\frac{\lambda t}{v}} \left(1 - e^{-\frac{2\lambda b}{v}}\right)^{-1} \mathbb{I}_{\{0 \leq t \leq 2b\}}.$$

Therefore,

$$\mathbf{E}[r_i] = \mathbf{E}\left[\int_{u_i}^{u_i + v_i} f(vt)\, dt\right]$$

$$= \frac{\lambda}{2bv\left(1-e^{-\frac{2\lambda b}{v}}\right)} \int_0^{2b} dx \int_0^{2b} dp\, e^{-\frac{\lambda p}{v}} \int_x^{x+p} dt f\left(S_1\left(t\right)\right)$$

$$= -\frac{1}{2b} \int_0^{2b} dx \int_x^{x+2b} f\left(S_k\left(t\right)\right) dt \frac{e^{-\frac{2\lambda b}{v}}}{1-e^{-\frac{2\lambda b}{v}}}$$

$$+ \frac{1}{2b\left(1-e^{-\frac{2\lambda b}{v}}\right)} \int_0^{2b} dx \int_0^{2b} dp f\left(S_1\left(x+p\right)\right) e^{-\frac{\lambda p}{v}}$$

$$= -\frac{2e^{-\frac{2\lambda b}{v}}}{\left(1-e^{-\frac{2\lambda b}{v}}\right)v} \int_0^b f\left(x\right) dx + \frac{1}{\lambda b} \int_0^b f\left(x\right) dx. \qquad [2.19]$$

The strong law of large numbers for $\{r_i,\ i \geq 1\}$ implies

$$\lim_{N\to+\infty} \frac{1}{N} \sum_{i=1}^{N} r_i = \mathbf{E}\left[r_i\right] \quad (a.s.).$$

Since $\theta_j = s_{j+1} - s_j,\ j = 1, 2, \ldots$ are independent exponentially distributed random variables, we have the following strong law of large numbers:

$$\lim_{N\to+\infty} \frac{1}{N} \sum_{j=1}^{N} \left[\frac{\left(s_j - s_{j-1}\right) v}{2b}\right] = \mathbf{E}\left[\frac{\theta_j v}{2b}\right]$$

$$= \sum_{n=1}^{\infty} n \left(e^{-\frac{2n\lambda b}{v}} - e^{-\frac{2(n+1)\lambda b}{v}}\right) = \frac{e^{-\frac{2\lambda b}{v}}}{1-e^{-\frac{2\lambda b}{v}}}. \qquad [2.20]$$

Combining equations [2.17] to [2.20], we get

$$P\left(\lim_{T\to+\infty} \frac{1}{T} \int_0^T f\left(S_1\left(t\right)\right) dt\right.$$

$$\left. = \lim_{T\to+\infty} \frac{N\left(t\right)}{T} \frac{1}{N\left(T\right)} \left(\sum_{j=1}^{N(T)} \int_{s_{j-1}}^{s_j} f\left(S_1\left(t\right)\right) dt\right) = \frac{1}{b} \int_0^b f\left(x\right) dx\right) = 1.$$

This concludes the proof.

Therefore, the limiting distribution of $S_k\left(t\right)$ as $t \to +\infty$ for all $k = 1, \ldots, n$ is the uniform distribution on $\left[0, b\right]$.

LEMMA 2.3.– Suppose that the initial distribution of a telegraph particle with reflecting boundaries 0 and b is uniform on $[0, b]$. Then it remains uniformly distributed for all $t > 0$.

PROOF.– Denote by $p(t,x)$ the probability density of this particle's position at time t.

Now, let us denote by $p(t,x \mid y_k)$ the probability density of the process $S_k(t)$ and its position at time t. It was shown in Masoliver *et al.* (1993a) that for $x \in [0,b]$

$$p(t,x \mid y_k) = \frac{1}{b}$$
$$+\frac{2}{b}e^{-\lambda t}\sum_{n=1}^{\infty}\left\{\left[\cosh(\theta_n t) + \frac{\lambda}{\theta_n}\sinh\theta_n t\right]\cos\left(\frac{\pi n y_k}{b}\right)\cos\left(\frac{\pi n x}{b}\right)\right\},$$

where

$$\theta_n = \left(\lambda^2 - \frac{\pi^2 v^2}{b^2}n^2\right)^{1/2}.$$

It is easily seen that for any $t > 0$ and $x \in [0,b]$,

$$p(t,x) = \frac{1}{b}\int_0^b p(t,x \mid y_k)dy_k = \frac{1}{b},$$

and this concludes the proof.

Now, let us consider the system of processes $\overline{S}_k(t)$ with the limiting distribution of the respective processes $S_k(t)$, $k = 1,2,\dots n$. That is, according to theorem 2.3 and lemma 2.3 for each $t \geq 0$, $\overline{S}_k(t)$, $k = 1,2,\dots n$ are independent identically distributed with the uniform distribution on $[0,b]$.

For this case, the particle positions $x_k(t)$, $k = 1,2,\dots,n$ satisfy the following equations of order statistics $x(y_1,t) = \overline{S}_{(1)}(t)$, $x(y_2,t) = \overline{S}_{(2)}(t)$, ..., $x(y_n,t) = \overline{S}_{(n)}(t)$.

Consider the following function:

$$p(x) = \begin{cases} \frac{x}{b}, & x \in [0,b] \\ 0, & x \notin [0,b]. \end{cases}$$

Hence, the distributions $\pi_k(\cdot)$ of the particle positions $x_k(t)$, $k \in \{1,2,\dots,n\}$ are as follows:

$$\pi_k(x) = P(x_k(t) \leq x) = I_{p(x)}(k, n-k+1),$$

where

$$I_{p(x)}(k,n-k+1) = \frac{\int_0^{p(x)} t^{k-1}(1-t)^{n-k}dt}{\int_0^1 t^{k-1}(1-t)^{n-k}dt}.$$

Let us study the number of collisions $C_{(1,2,\dots,n)}(0,t)$ of particles $x_k(t)$, $k = 1,2,\dots,n$ during the time interval $(0,t)$.

THEOREM 2.4.– The expectation of $C_{(1,2,\dots,n)}(0,t)$ has the form

$$\mathbf{E}\left[C_{(1,2,\dots,n)}(0,t)\right] = \frac{n(n-1)}{2}ct, \quad where \quad \frac{v}{b} \leq c \leq \frac{4v}{b}.$$

PROOF.– It is easy to see that $C_{(1,2,\dots,n)}(0,t)$ is the number of intersections of $\overline{S}_k(t)$, $k = 1, 2, \dots n$ up to time t.

Denote by $I_{(k,l)}(0,t)$, $k \neq l$ the number of intersections of processes $\overline{S}_k(t)$ and $\overline{S}_l(t)$ during $(0,t)$.

Then,

$$C_{(1,2,\dots,n)}(0,t) = \sum_{1\leq k<l\leq n} I_{(k,l)}(0,t).$$

Therefore, let us analyze the distribution of $I_{(k,l)}(0,t)$, $k \neq l$.

Since $\overline{S}_k(t)$ and $\overline{S}_l(t)$ have the distribution $p(x) = \frac{x}{b}$ on $[0,b]$, the probability of their intersections $I_{(k,l)}(t, t+\Delta t)$ during $(t, t+\Delta t)$ satisfies the following inequalities for $x \in (a,b)$:

$$\frac{1}{4}P\left(\left|\overline{S}_k(t) - \overline{S}_l(t)\right| \leq 2\Delta tv\right)e^{-2\lambda\Delta t} \leq P\left(I_{(k,l)}(t,t+\Delta t) \geq 1\right)$$
$$\leq P\left(\left|\overline{S}_k(t) - \overline{S}_l(t)\right| \leq 2\Delta tv\right). \qquad [2.21]$$

Hence, since

$$P\left(\left|\overline{S}_k(t) - \overline{S}_l(t)\right| \leq 2\Delta tv\right) = O(\Delta t)$$

and

$$\frac{1}{4}P\left(\left|\overline{S}_k(t) - \overline{S}_l(t)\right| \leq 2\Delta tv\right)e^{-2\lambda\Delta t} = O(\Delta t),$$

we have that

$$P\left(I_{(k,l)}(t,t+\Delta t) \geq 1\right) = O(\Delta t). \qquad [2.22]$$

For $n \geq 2$, we have

$$P(I_{(k,l)}(t,t+\Delta t) = n) \leq P(\left|\overline{S}_k(t) - \overline{S}_l(t)\right| \leq 2\Delta tv)\left(1 - e^{-\lambda\Delta t}\right)^{2(n-1)}$$
$$+P\left(\left\{\overline{S}_k(t), \overline{S}_l(t) \in [0, 2\Delta tv]\right\} \cup \left\{\overline{S}_k(t), \overline{S}_l(t) \in [b - 2\Delta tv, b]\right\}\right)$$

$$\times\left(1-e^{-\lambda\Delta t}\right)^{(n-1)} \leq \frac{4\Delta tv}{b}(\lambda\Delta t)^{2(n-1)} + 2\frac{(2\Delta tv)^2}{b^2}(\lambda\Delta t)^{(n-1)}. \quad [2.23]$$

Taking into account equations [2.22] and [2.23], we conclude that

$$P\left(I_{(k,l)}\left(t, t+\Delta t\right) = 1\right) = O\left(\Delta t\right).$$

Therefore, for $x \in (a, b)$

$$O\left(\Delta t\right) = \frac{1}{4}P\left(\left|\overline{S}_k\left(t\right) - \overline{S}_l\left(t\right)\right| \leq 2\Delta tv\right)e^{-2\lambda\Delta t}$$

$$\leq \mathbf{E}\left[I_{(k,l)}\left(t, t+\Delta t\right)\right] \leq P\left(\left|\overline{S}_k\left(t\right) - \overline{S}_l\left(t\right)\right| \leq 2\Delta tv\right)$$

$$+\frac{4\Delta tv}{b}\sum_{n\geq 1} n\left((\lambda\Delta t)^{2(n-1)} + \frac{2\Delta tv}{b}(\lambda\Delta t)^{(n-1)}\right) = \frac{4\Delta tv}{b} + o\left(\Delta t\right). \quad [2.24]$$

We have the additive property of $\mathbf{E}\left[I_{(k,l)}\left(t_1, t_2\right)\right]$: for any $s \in (t_1, t_2)$

$$\mathbf{E}\left[I_{(k,l)}\left(t_1, t_2\right)\right] = \mathbf{E}\left[I_{(k,l)}\left(t_1, s\right)\right] + \mathbf{E}\left[I_{(k,l)}\left(s, t_2\right)\right].$$

Hence, there exists a constant $c > 0$ such that

$$\mathbf{E}\left[I_{(k,l)}\left(0, t\right)\right] = ct.$$

This implies that

$$\mathbf{E}\left[C_{(1,2,\ldots,n)}\left(0, t\right)\right] = \frac{n\left(n-1\right)}{2}ct. \quad [2.25]$$

It follows from [2.21] and [2.24] that $\frac{v}{b} \leq c \leq \frac{4v}{b}$.

2.1.5. *Estimation of the number of particle collisions for systems without boundaries*

Now, we will focus on finding the asymptotic estimate of the number of particle collisions as time goes to infinity. Unlike the analysis in section 2.5.4, in this case, the system of particles does not have reflective boundaries and the particles can move arbitrarily far away in a straight line.

Assume that $y_1 < y_2 < \ldots < y_n$ and consider n independent telegraph processes

$$\frac{d}{dt}S_i\left(t\right) = v(-1)^{\xi_i(t)},\ i = 1, 2, \ldots, n$$

with the initial condition $S_i(0) = y_i$. As defined before, $\{\xi_i(t), t \geq 0, i = 1, \ldots, n\}$ are independent Markov processes on the phase space $\{0, 1\}$ with infinitesimal matrix

$$Q = \lambda \begin{pmatrix} -1 & 1 \\ 1 & -1 \end{pmatrix}.$$

We assume that all processes have absolute velocity v and a parameter of switching processes λ. In addition, starting from y_i the process $S_i(t)$ has equal probabilities of initial directions of motion.

Denote by $x(y_i, t)$ the position at time t of particle i that started from y_i. As mentioned in section 2.5.4, particle i moves according to the telegraph process $S_i(t)$ before its first elastic collision with another particle. At the instant of collision, the particles exchange momentum or, in other words, they exchange telegraph processes that describe their movement.

As before, the particle trajectories $x(y_i, t)$, $i = 1, 2, \ldots,$ are the order statistics of the processes $S_i(t)$, $i = 1, 2, \ldots$, i.e.

$$x(y_1, t) = S_{(1)}(t),$$

$$x(y_2, t) = S_{(2)}(t), \ldots, x(y_n, t) = S_{(n)}(t).$$

Let us define $\boldsymbol{\eta}(t) = (\xi_1(t), \xi_2(t), \ldots, \xi_n(t))$. We also define the number of collisions of particles with trajectories $x(y_i, t)$, $i = 1, 2, \ldots, n$ in the time interval $(0, t)$, $t > 0$, as $C_{(k_1, k_2, \ldots, k_n)}(t, y_1, y_2, \ldots, y_n)$, $k_i \in \{0, 1\}$, when $\boldsymbol{\eta}(0) = (k_1, k_2, \ldots, k_n)$.

Let us consider the mean value

$$H_{(k_1, k_2, \ldots, k_n)}(t, y_1, y_2, \ldots, y_n) = \mathbf{E}\left[C_{(k_1, k_2, \ldots, k_n)}(t, y_1, y_2, \ldots, y_n)\right].$$

THEOREM 2.5.– The mean number of collisions follows the asymptotic

$$H_{(k_1, k_2, \ldots, k_n)}(t, y_1, y_2, \ldots, y_n) \sim \frac{n(n-1)\sqrt{2\lambda}}{\sqrt{\pi}} \sqrt{t}, \quad t \to \infty.$$

PROOF.– Let us denote as $I_{(k_1, k_2, \ldots, k_n)}(t, y_1, y_2, \ldots, y_n)$ the number of collisions of particles or, equivalently, the number of intersections of processes $S_i(t)$, $i = 1, 2, \ldots, n$ during time interval $(0, t)$, $t > 0$, i.e.

$$C_{(k_1, k_2, \ldots, k_n)}(t, y_1, y_2, \ldots, y_n) = I_{(k_1, k_2, \ldots, k_n)}(t, y_1, y_2, \ldots, y_n),$$

It is easily seen that the number of intersections of all processes is equal to the sum of intersections of different pairs of these processes

$$I_{(k_1,k_2,\ldots,k_n)}(t,y_1,y_2,\ldots,y_n)=\sum_{1\le i<j\le n} I_{(k_i,k_j)}(t,y_j-y_i),$$

where $I_{(k_i,k_j)}(t,y_j-y_i)$ is the number of intersections of processes $S_i(t)$ and $S_j(t)$.

Hence, the mean number of intersections is given by

$$H_{(k_1,k_2,\ldots,k_n)}(t,y_1,y_2,\ldots,y_n)=\mathbf{E}\left[C_{(k_1,k_2,\ldots,k_n)}(t,y_1,y_2,\ldots,y_n)\right]$$

$$=\sum_{1\le i<j\le n}\mathbf{E}\left[I_{(k_i,k_j)}(t,y_j-y_i)\right].$$

Therefore, to estimate $H_{(k_1,k_2,\ldots,k_n)}(t,y_1,y_2,\ldots,y_n)$, it is enough to find the estimation of $\mathbf{E}\left[I_{(k_i,k_j)}(t,z)\right]$, where $z=y_j-y_i$.

Denote by τ_i, $i=1,2,3\ldots$, the instants of collision occurrences of any couple of particles, say particles 1, 2, and let us define $v_i=\tau_{i+1}-\tau_i$, $i\ge 1$, and $v_0=\tau_1$.

From equation [2.7], f or $z=0$, the Laplace transform of v_i, for $i\ge 1$ is of the form

$$\mathbf{E}\left[e^{-sv_i}\right]=\frac{s+2\lambda-\sqrt{s^2+4\lambda s}}{s+2\lambda+\sqrt{s^2+4\lambda s}},\qquad s>0.$$

By using Bateman (1955, no. 49, p. 237), we have

$$f_{(1,0)}(t,0)=\mathcal{L}^{-1}\left(\frac{s+2\lambda-\sqrt{s^2+4\lambda s}}{s+2\lambda+\sqrt{s^2+4\lambda s}},t\right)$$

$$=\frac{1}{4\lambda^2}\mathcal{L}^{-1}\left(\left(s+2\lambda-\sqrt{s^2+4\lambda s}\right)^2,t\right)$$

$$=e^{-2\lambda t}\int_0^t\frac{I_1(2\lambda(t-v))}{(t-v)}\frac{I_1(2\lambda v)}{v}dv=\frac{2e^{-2\lambda t}}{t}I_2(2\lambda t).$$

Therefore, the distribution function is given by

$$F_{(1,0)}(x)=\int_0^x f_{(1,0)}(t,0)\,dt=\int_0^x\frac{2e^{-2\lambda t}}{t}I_2(2\lambda t)\,dt.$$

Denote by

$$F_{(k_1,k_2)}(x,z)=\int_0^x f_{(k_1,k_2)}(t,z)\,dt.$$

It is well known (Feller 1968) that if

$$1 - F_{(1,0)}(x) = x^{-\alpha} L(x),$$

where $0 < \alpha < 1$, and $L(x)$ is a slowly varying function, then

$$H_{(k_1,k_2)}(t,z) \sim \frac{\sin(\alpha\pi)}{\alpha\pi} \frac{t^{\alpha}}{L(t)}.$$

Consider

$$\lim_{x\to\infty} \left(1 - F_{(1,0)}(x)\right)\sqrt{x} = \lim_{x\to\infty} \left(1 - \int_0^x \frac{2e^{-2\lambda t}}{t} I_2(2\lambda t)\, dt\right)\sqrt{x}$$

$$= \lim_{x\to\infty} e^{-2\lambda x}\sqrt{x} I_2(2\lambda x) = \frac{1}{\sqrt{2\pi\lambda}}.$$

It is easily seen that for $s > 0$

$$\lim_{x\to\infty} \left(1 - F_{(1,0)}(sx)\right)\sqrt{sx} = \lim_{x\to\infty} e^{-2\lambda sx}\sqrt{sx} I_2(2\lambda sx) = \frac{1}{\sqrt{2\pi\lambda}}.$$

Therefore, $L(x) = \left(1 - F_{(1,0)}(x)\right)\sqrt{x}$ is a slowly varying function and

$$1 - F_{(1,0)}(x) = x^{-1/2} L(x).$$

Hence,

$$H_{(k_1,k_2)}(t,z) \sim \frac{2\sqrt{2\lambda}}{\sqrt{\pi}}\sqrt{t}, \quad t \to \infty.$$

This implies

$$H_{(k_1,k_2,\ldots,k_n)}(t, y_1, y_2, \ldots, y_n) = \sum_{1\le i<j\le n} \mathbf{E}\left[I_{(k_i,k_j)}(t, y_j - y_i)\right]$$

$$\sim \frac{n(n-1)\sqrt{2\lambda}}{\sqrt{\pi}}\sqrt{t}, \quad as\ t \to \infty. \qquad [2.26]$$

Comparing [2.24] and [2.26], we can see that, in the case of the absence of reflective boundaries, the number of particle collisions grows more slowly (in proportion $\sqrt{t}$). This means that the system without reflecting boundaries expends the energy on its expansion.

2.2. Interaction of particles governed by generalized integrated telegraph processes: a semi-Markov case

Many contributions on telegraph random processes, and some of their variations, have been published; however most of these contributions are devoted to describing the random motion of a single particle in a Markov or semi-Markov medium; see De Gregorio (2012) and Pogorui and Rodríguez-Dagnino (2013) for two recent works in this direction. In this section, we are dealing with the interaction or collision of two or more particles that switch their position according to a semi-Markov process. The resulting process is a generalization of the well-known telegraph process, and it is called the *generalized integrated telegraph process* since it has semi-Markov switching times.

We construct a set of processes with interactions where the particles move according to telegraph processes. Our model is different from the model presented in Chapter 6 of Kornfeld *et al.* (1982), where the particles move in straight lines with constant velocity up to collision. The motion in such a case can be seen as trajectories of a billiard in the n-dimensional simplex. Our model cannot be reduced to a billiard type of system. Another similar model of interacting particles was studied in Konarovskii (2011), as one of the representatives of the Arratia flow (Arratia 1979). In such a diffusion model, the particles stick together under collision and change the diffusion coefficient at the moment of coalescence.

Coalescence is an important phenomenon considered in several applications. For instance, customer coalescence in queueing networks (Economou 2002), to model surface cracks in materials (Ignatovich *et al.* 2004), and as remarked by Marcus: "for the growth of cloud droplets into raindrops, the formation of molecular polymers and the growth of particles in a colloidal suspension" (Marcus 1968), just to mention a few.

Let $\{\xi(t),\, t \geq 0\}$ be a semi-Markov alternating process on the phase space $\{0, 1\}$ and having a semi-Markov kernel (Korolyuk and Korolyuk 1999),

$$Q(t) = G(t) \begin{pmatrix} 0 & 1 \\ 1 & 0 \end{pmatrix},$$

where $G(t)$ is a general distribution function.

Let us define as $\{S(t)\}$ the telegraph process with semi-Markov switching times when

$$S(t) = y_0 + v \int_0^t (-1)^{\xi(s)}\, ds, \qquad v = \text{constant} > 0.$$

Then, the process $\xi(t)$ is called the semi-Markov switching process for the generalized integrated telegraph process $S(t)$.

For instance, for the exponential distribution $G(t) = (1-e^{-\lambda t})\,\mathbb{I}_{\{t\geq 0\}}$, the process $S(t)$ reduces to the well-known telegraph process (Korolyuk and Korolyuk 1999).

For an arbitrary finite set of ordered real numbers $y_1 < y_2 < \cdots < y_n$, we consider a family of independent telegraph processes $S_i(t)$, $t \geq 0$, such that $S_i(0) = y_i$ for $i = 1, 2, \ldots, n$ with absolute velocity v and intensity or parameter of switching process λ. A random time between two consecutive switching instants is distributed with cdf $G(t)$ and pdf $g(t)$.

Denote by $x(y_i, t)$ the position at time t of the particle i, which starts from site y_i. Suppose that such a particle evolves as a telegraph process $S_i(t)$ up to the hard collision with another particle.

The meaning of *hard* collision of two particles is that at the time of the collision the particles change to the opposite direction, i.e. the particles exchange the telegraph processes that describe their movement.

It is easily verified that in the case of hard collisions the trajectories of the particles $x(y_i, t), i = 1, 2, \ldots$ coincide with the order statistics of $S_i(t), i = 1, 2, \ldots$ as follows:

$$x(y_1, t) = S_{(1)}(t),\ x(y_2, t) = S_{(2)}(t),\ \ldots,\ x(y_n, t) = S_{(n)}(t). \qquad [2.27]$$

REMARK 2.3.– *It follows from the description of $x(y_i, t)$ that $x(y_1, t) \leq x(y_2, t) \leq \cdots \leq x(y_n, t)$ for any $t \in [0, T]$. Models achieving these kinds of conditions for Wiener processes with several types of coalescence after* hard *collision are called Arratia flows, and they have been studied in Arratia (1979); Dorogovtsev (2007a, 2010); Le and Raimond (2004); Tóth and Werener (1998), and others.*

2.2.1. *Laplace transform of the distribution of the first collision of two particles*

Let us consider two particles, say 1 and 2, on the line, where each particle can move in two opposite directions. Starting at position $x_i \in \mathbb{R}$, $i = 1, 2$, particle i moves at constant velocity $v > 0$ in one of the two directions during a random time interval that we assume as general distributed with cdf $G(t)$ and pdf $g(t)$. Thus, such a particle changes its direction and so on. We will say, in the rest of the section, that such a particle is a generated telegrapher particle since its motion satisfies the generated telegrapher's process. As stated above, the particle satisfies the telegraph equation in the case of an exponential distribution for $G(t)$ (Kac 1974); see also the seminal work on the telegraph equation by Kelvin (1854).

We will denote as $\{\xi_\ell(t)\}$, $\ell = 1, 2$ the alternating two-state semi-Markov process with phase space $\{0, 1\}$ and semi-Markov kernel (Korolyuk and Korolyuk 1999),

$$Q(t) = G(t)\begin{pmatrix} 0 & 1 \\ 1 & 0 \end{pmatrix}.$$

Let $x_i(t)$ be the position of particle i at time t, $t \geq 0$. Thus, it is easily seen that

$$\frac{d}{dt}\, x_i(t) = v\,(-1)^{\xi_i(t)},$$

with the initial condition $x_i(0) = x_i$.

Now, let us define $z = (x_2 - x_1) > 0$, $\Delta(t) = x_2(t) - x_1(t)$, and the two-dimensional process $\eta(t) = (\xi_1(t), \xi_2(t))$. Then, we consider

$$\tau_{(k_1,k_2)}(z) = \inf\{t \geq 0 \,:\, \Delta(t) = 0,\ \eta(0) = (k_1, k_2)\}, \quad k_j \in \{0, 1\},$$

as the instant of the first particle collision assuming $\eta(0) = (k_1, k_2)$.

Our purpose in this section is to calculate the probability distribution of the random variable $\tau_{(k_1,k_2)}(z)$.

We will calculate the Laplace transform of the random variable $\tau_{(k_1,k_2)}(z)$

$$\varphi_{(k_1,k_2)}(s, z) = \mathbf{E}\left[e^{-s\tau_{(k_1,k_2)}(z)}\right], \quad s > 0, \qquad [2.28]$$

for each of the four cases $(k_1, k_2) \in \{0, 1\} \times \{0, 1\}$.

By using similar arguments to those applied to obtain the integral equation in renewal theory, we can obtain a system of four integral equations for these Laplace transforms. For instance, we should consider the case when $\eta(0) = (0, 1)$. In this case, particles move in opposite directions to meet each other. If renewal instants for both particles do not occur during time $\frac{z}{2v}$, then the particles collide at time $t = \frac{z}{2v}$. Therefore, we have Laplace transform $e^{-\frac{sz}{v}}\left(1 - G\left(\frac{z}{2v}\right)\right)^2$. If the first renewal occurs at time $\frac{u}{2v}$, $u < z$, then from this instant the collision time of two particles is given by $\tau_{(1,1)}(z - u)$ or $\tau_{(0,0)}(z - u)$ depending on whether $\xi_1(t)$ or $\xi_2(t)$ has changed its state at time $\frac{u}{2v}$. For the rest of the cases, the renewal arguments are similar.

Hence, these Laplace transforms [2.28] satisfy a system of four integral equations as follows:

$$\varphi_{(0,1)}(s, z) = e^{-\frac{sz}{v}}\left(1 - G\left(\frac{z}{2v}\right)\right)^2 + \frac{1}{2v}\int_0^z e^{-\frac{su}{2v}} M\left(\frac{u}{2v}\right)$$

$$\times \left(\varphi_{(0,0)}(s, z - u) + \varphi_{(1,1)}(s, z - u)\right) du,$$

where $M(t) := g(t)(1 - G(t))$, $t \geq 0$. Equivalently,

$$\varphi_{(0,1)}(s, z) = e^{-\frac{sz}{v}}\left(1 - G\left(\frac{z}{2v}\right)\right)^2 + \frac{1}{2v}\int_0^z e^{-\frac{s(z-u)}{2v}} M\left(\frac{z - u}{2v}\right)$$

$$\times \left(\varphi_{(0,0)}(s,u) + \varphi_{(1,1)}(s,u)\right) du,$$

and

$$\varphi_{(1,0)}(s,z) = \frac{1}{2v}\int_0^{\infty} e^{-\frac{su}{2v}} M\left(\frac{u}{2v}\right)\left(\varphi_{(0,0)}(s,z+u) + \varphi_{(1,1)}(s,z+u)\right) du.$$

Due to the symmetry of the problem $\varphi_{(0,0)}(s,z) = \varphi_{(1,1)}(s,z)$, i.e. absolute velocity is the same for both particles, and the direction in which they move does not matter. For instance, $\varphi_{(0,0)}(s,z)$ is related to particles that move to the right, whereas the case $\varphi_{(1,1)}(s,z)$ corresponds to particles that move to the left. Thus, it is easily seen that

$$\varphi_{(0,0)}(s,z) = \varphi_{(1,1)}(s,z)$$
$$= \int_0^{\infty} M(u)\, e^{-su}\, du\, \left(\varphi_{(0,1)}(s,z) + \varphi_{(1,0)}(s,z)\right)$$

then

$$\varphi_{(0,0)}(s,z) + \varphi_{(1,1)}(s,z)$$
$$= 2\int_0^{\infty} M(u)\, e^{-su}\, du\, (\varphi_{(0,1)}(s,z) + \varphi_{(1,0)}(s,z)). \qquad [2.29]$$

Hence,

$$\varphi_{(0,1)}(s,z) = e^{-\frac{sz}{2v}}\left(1 - G\left(\frac{z}{2v}\right)\right)^2 + \frac{1}{v} e^{-\frac{sz}{2v}} \int_0^{\infty} M(x)\, e^{-sx}\, dx$$
$$\times \int_0^{z} M\left(\frac{z-u}{2v}\right) e^{\frac{su}{2v}} \left(\varphi_{(0,1)}(s,u) + \varphi_{(1,0)}(s,u)\right) du, \qquad [2.30]$$

and

$$\varphi_{(1,0)}(s,z) = \frac{1}{v} e^{\frac{sz}{2v}} \int_0^{\infty} M(x)\, e^{-sx}\, dx$$
$$\times \int_z^{\infty} M\left(\frac{u-z}{2v}\right) e^{-\frac{su}{2v}} \left(\varphi_{(0,1)}(s,u) + \varphi_{(1,0)}(s,u)\right) du. \qquad [2.31]$$

It follows that from equations [2.30] to [2.31], we obtain

$$\varphi_{(0,1)}(s,z) + \varphi_{(1,0)}(s,z) = e^{-\frac{sz}{2v}}\left(1 - G\left(\frac{z}{2v}\right)\right)^2 + \frac{1}{v}\int_0^{\infty} M(x)\, e^{-sx}\, dx$$
$$\times \int_0^{\infty} M\left(\frac{|z-u|}{2v}\right) e^{-\frac{s(|z-u|)}{2v}} \left(\varphi_{(0,1)}(s,u) + \varphi_{(1,0)}(s,u)\right) du. \qquad [2.32]$$

Suppose that we have solved equation [2.32] for the unknown $\varphi_{(0,1)}(s,z) + \varphi_{(1,0)}(s,z)$. Thus, by substituting $\varphi_{(0,1)}(s,z) + \varphi_{(1,0)}(s,z)$ into equations [2.30] to [2.31], we obtain $\varphi_{(0,1)}(s,z)$ and $\varphi_{(1,0)}(s,z)$.

2.2.2. *Semi-Markov case*

Consider the case where $G(z) = 1 - \sqrt{2e^{-\lambda z} - e^{-2\lambda z}}$. This is a non-exponential example to show our method for semi-Markov processes. In this case, we have

$$M(z) = \lambda\left(e^{-\lambda z} - e^{-2\lambda z}\right), \quad g(z) = \frac{\lambda\left(1 - e^{-\lambda z}\right)}{\sqrt{2e^{\lambda z} - 1}},$$

$$(1 - G(z))^2 = 2e^{-\lambda z} - e^{-2\lambda z}.$$

Then,

$$\frac{1}{v}\int_0^\infty M(x)e^{-sx}\,dx = \frac{\lambda}{v}\int_0^\infty \left(e^{-\lambda z} - e^{-2\lambda z}\right)e^{-sx}dx = \frac{\lambda^2}{v\left(2\lambda^2 + 3\lambda s + s^2\right)}.$$

By using these results in equation [2.30], we obtain

$$\varphi_{(0,1)}(s,z) = \left(2e^{-\frac{s+\lambda}{2v}z} - e^{-\frac{s+2\lambda}{2v}z}\right) + \frac{\lambda^2}{v\left(2\lambda^2 + 3\lambda s + s^2\right)}$$
$$\times \int_0^z \lambda\left(e^{-\frac{s+\lambda}{2v}(z-u)} - e^{-\frac{s+2\lambda}{2v}(z-u)}\right)\left(\varphi_{(0,1)}(s,u) + \varphi_{(1,0)}(s,u)\right)du. \qquad [2.33]$$

Now, we will find an equivalent partial differential equation (PDE) to equation [2.33]. Then, the derivative of equation [2.33] can be written as

$$\frac{\partial}{\partial z}\varphi_{(0,1)}(s,z) = -\frac{s+\lambda}{v}e^{-\frac{s+\lambda}{2v}z} + \frac{s+2\lambda}{2v}e^{-\frac{s+2\lambda}{2v}z}$$
$$-\frac{\lambda^3}{2v^2(s+2\lambda)}\int_0^z \left(e^{-\frac{s+\lambda}{2v}(z-u)}\right)\left(\varphi_{(0,1)}(s,u) + \varphi_{(1,0)}(s,u)\right)du$$
$$+\frac{\lambda^3}{2v^2(s+\lambda)}\int_0^z \left(e^{-\frac{s+2\lambda}{2v}(z-u)}\right)\left(\varphi_{(0,1)}(s,u) + \varphi_{(1,0)}(s,u)\right)du. \qquad [2.34]$$

After multiplying equation [2.33] by $\frac{s+2\lambda}{2v}$ and adding this result to equation [2.34], we obtain

$$\frac{\partial}{\partial z}\varphi_{(0,1)}(s,z) + \frac{s+2\lambda}{2v}\varphi_{(0,1)}(s,z) = \frac{\lambda}{v}e^{-\frac{s+\lambda}{2v}z}$$
$$+\frac{\lambda^4}{2v^2\left(2\lambda^2 + 3\lambda s + s^2\right)}\int_0^z e^{-\frac{s+\lambda}{2v}(z-u)}\left(\varphi_{(0,1)}(s,u) + \varphi_{(1,0)}(s,u)\right)du. \qquad [2.35]$$

Then, by taking the derivative of equation [2.35], and identifying that one part of this equation is equation [2.35], we have

$$\frac{\partial^2}{\partial z^2}\varphi_{(0,1)}(s,z)+\frac{s+2\lambda}{2v}\frac{\partial}{\partial z}\varphi_{(0,1)}(s,z)$$

$$=-\frac{s+\lambda}{2v}\left(\frac{\partial}{\partial z}\varphi_{(0,1)}(s,z)+\frac{s+2\lambda}{2v}\varphi_{(0,1)}(s,z)\right)$$

$$+\frac{\lambda^4\left(\varphi_{(0,1)}(s,u)+\varphi_{(1,0)}(s,u)\right)}{2v^2\left(2\lambda^2+3\lambda s+s^2\right)}.$$

By making some algebraic simplifications, we finally obtain the equivalent second-order PDE

$$\frac{\partial^2}{\partial z^2}\varphi_{(0,1)}(s,z)+\frac{2s+3\lambda}{2v}\frac{\partial}{\partial z}\varphi_{(0,1)}(s,z)+\frac{2\lambda^2+3\lambda s+s^2}{4v^2}\varphi_{(0,1)}(s,z)$$

$$-\frac{\lambda^4\left(\varphi_{(0,1)}(s,u)+\varphi_{(1,0)}(s,u)\right)}{2v^2\left(2\lambda^2+3\lambda s+s^2\right)}=0. \qquad [2.36]$$

Similarly, for $\varphi_{(1,0)}(s,z)$, we have

$$\varphi_{(1,0)}(s,z)=\frac{\lambda^3}{v\left(2\lambda^2+3\lambda s+s^2\right)}\int_z^\infty\left(e^{-\frac{s+\lambda}{2v}(u-z)}-e^{-\frac{s+2\lambda}{2v}(u-z)}\right)$$

$$\times\left(\varphi_{(0,1)}(s,u)+\varphi_{(1,0)}(s,u)\right)du.$$

The first derivative is given by

$$\frac{\partial}{\partial z}\varphi_{(1,0)}(s,z)$$

$$=\frac{\lambda^3}{2v^2\left(s+2\lambda\right)}\int_z^\infty\left(e^{-\frac{s+\lambda}{2v}(u-z)}\right)\left(\varphi_{(0,1)}(s,u)+\varphi_{(1,0)}(s,u)\right)du$$

$$-\frac{\lambda^3}{2v^2\left(s+\lambda\right)}\int_z^\infty\left(e^{-\frac{s+2\lambda}{2v}(u-z)}\right)\left(\varphi_{(0,1)}(s,u)+\varphi_{(1,0)}(s,u)\right)du.$$

Therefore,

$$\frac{\partial}{\partial z}\varphi_{(1,0)}(s,z)-\frac{s+\lambda}{2v}\varphi_{(1,0)}(s,z)=0$$

$$=-\frac{\lambda^4}{2v^2\left(2\lambda^2+3\lambda s+s^2\right)}\int_z^\infty e^{-\frac{s+2\lambda}{2v}(u-z)}\left(\varphi_{(0,1)}(s,u)+\varphi_{(1,0)}(s,u)\right)du.$$

Hence,

$$\frac{\partial^2}{\partial z^2}\varphi_{(1,0)}(s,z) - \frac{s+\lambda}{2v}\frac{\partial}{\partial z}\varphi_{(1,0)}(s,z)$$

$$= \frac{s+2\lambda}{2v}\left(\frac{\partial}{\partial z}\varphi_{(1,0)}(s,z) - \frac{s+\lambda}{2v}\varphi_{(1,0)}(s,z)\right)$$

$$+\frac{\lambda^4}{2v^2\left(2\lambda^2+3\lambda s+s^2\right)}\left(\varphi_{(0,1)}(s,z)+\varphi_{(1,0)}(s,z)\right)$$

and we obtain the equivalent second-order PDE

$$\frac{\partial^2}{\partial z^2}\varphi_{(1,0)}(s,z) - \frac{2s+3\lambda}{2v}\frac{\partial}{\partial z}\varphi_{(1,0)}(s,z) + \frac{2\lambda^2+3\lambda s+s^2}{4v^2}\varphi_{(1,0)}(s,z)$$

$$-\frac{\lambda^4}{2v^2\left(2\lambda^2+3\lambda s+s^2\right)}\left(\varphi_{(0,1)}(s,z)+\varphi_{(1,0)}(s,z)\right) = 0. \qquad [2.37]$$

Thus, we will find the solution of the system of equations given by equations [2.36] and [2.37]. Let us consider the matrix M, which consists of the factors of this system of equations:

$$M = \begin{pmatrix} \frac{\partial^2}{\partial z^2} + \frac{2s+3\lambda}{2v}\frac{\partial}{\partial z} + \Lambda_s & -\frac{\lambda^4}{2v^2(2\lambda^2+3\lambda s+s^2)} \\ -\frac{\lambda^4}{2v^2(2\lambda^2+3\lambda s+s^2)} & \frac{\partial^2}{\partial z^2} - \frac{2s+3\lambda}{2v}\frac{\partial}{\partial z} + \Lambda_s \end{pmatrix},$$

where $\Lambda_s = \frac{\left(2\lambda^2+3\lambda s+s^2\right)^2-2\lambda^4}{4v^2(2\lambda^2+3\lambda s+s^2)}$. Let us calculate the determinant of the matrix M:

$$\det(M) = \frac{\partial^4}{\partial z^4} + \frac{2\left(2\lambda^2+3\lambda s+s^2\right)^2-4\lambda^4}{4v^2\left(2\lambda^2+3\lambda s+s^2\right)}\frac{\partial^2}{\partial z^2} - \frac{(2s+3\lambda)^2}{4v^2}\frac{\partial^2}{\partial z^2}$$

$$+\left(\frac{\left(2\lambda^2+3\lambda s+s^2\right)^2-2\lambda^4}{4v^2\left(2\lambda^2+3\lambda s+s^2\right)}\right)^2 - \left(\frac{\lambda^4}{2v^2\left(2\lambda^2+3\lambda s+s^2\right)}\right)^2.$$

After simplification, we have

$$\det(M) = \frac{\partial^4}{\partial z^4} - \frac{14\lambda^4+27\lambda^3 s+27\lambda^2 s^2+12\lambda s^3+2s^4}{4v^2\left(2\lambda^2+3\lambda s+s^2\right)}\frac{\partial^2}{\partial z^2}$$

$$+\frac{1}{16v^4}\left(2\lambda^2+3\lambda s+s^2\right)^2 - \frac{\lambda^4}{4v^4}$$

It is well known (Pinsky 1991) that the functions $\varphi_{(0,1)}(s,z)$ and $\varphi_{(1,0)}(s,z)$ satisfy the following equation:

$$\det(M)\, f = 0.$$

Consider it in more detail,

$$\frac{\partial^4}{\partial z^4} f(z) - \frac{14\lambda^4 + 27\lambda^3 s + 27\lambda^2 s^2 + 12\lambda s^3 + 2s^4}{4v^2\left(2\lambda^2 + 3\lambda s + s^2\right)} \frac{\partial^2}{\partial z^2} f(z)$$

$$+ \left[\frac{1}{16v^4}\left(2\lambda^2 + 3\lambda s + s^2\right)^2 - \frac{\lambda^4}{4v^4}\right] f(z) = 0.$$

The only solution that satisfies both conditions $\varphi_{(0,1)}(0, z) = \varphi_{(0,1)}(s, 0) = 1$ and $\lim_{s\to\infty} \varphi_{(0,1)}(s, z) = 0, \quad z > 0$ is the following:

$$\varphi_{(0,1)}(s, z) = \exp\left\{\frac{-z}{4v\left(2\lambda^2 + 3\lambda s + s^2\right)}\right.$$

$$\times\Big(\left(4 + 6s + 2s^2\right)\left(14\lambda^4 + 27\lambda^3 s + 27\lambda^2 s^2 + 12\lambda s^3 + 2s^4\right.$$

$$\left.\left. - \sqrt{196\lambda^8 + 564\lambda^7 s + 701\lambda^6 s^2 + 450\lambda^5 s^3 + 36\lambda^3 s^5 + 4\lambda^2 s^6}\right)\Big)^{\frac{1}{2}}\right\}. \qquad [2.38]$$

The expression for $\varphi_{(1,0)}(s, z)$ can be obtained from equation [2.31] in the following form:

$$\varphi_{(1,0)}(s, z) = a(s)\, \varphi_{(0,1)}(s, z),$$

where $a(s)$ is a function of s.

We could not calculate the inverse Laplace transform for $\varphi_{(0,1)}(s, z)$ but using this function, we can get some of the properties of the instant of the first collision of these particles $\tau_{(1,0)}(z)$.

Let us consider the following so-called Kac's condition (or the hydrodynamic limit): denote by $\lambda = \varepsilon^{-2}$, $v = \varepsilon^{-1}$, as $\varepsilon \to 0$, that is $v \to +\infty$, and $\lambda \to +\infty$, such that $\frac{v^2}{\lambda} \to 1$.

It is shown in Kac (1974) that the telegraph process under Kac's condition weakly converges to the Wiener process $W(t) \sim N(0, t)$.

It follows from equation [2.38] that for our case

$$\lim_{\varepsilon\to 0} \varphi_{(0,1)}(s, z) = e^{-\sqrt{\frac{3}{7}} z\sqrt{s}}.$$

Passing to the inverse Laplace transform, we have

$$f(t,z) = \mathcal{L}^{-1}\left(e^{-\sqrt{\frac{3}{7}}z\sqrt{s}}\right) = \frac{\sqrt{\frac{3}{7}}\, z \exp\left(-\frac{3z^2}{28t}\right)}{2\sqrt{\pi}t^{3/2}}.$$

It is well known that $f(t,z)$ is the pdf of a collision instant of two particles moving according to Wiener paths $W(t)$ with diffusion coefficient $\sigma^2 = \frac{3}{7}$, where $z > 0$ is the distance between the starting points of the particles. Denote this instant by $\tau_{(W)}(z)$

Therefore, $\tau_{(0,1)}(z)$ weakly converges to $\tau_{(W)}(z)$ under Kac's condition.

Next, the first moment can be obtained as

$$\mathbf{E}\left[\tau_{(0,1)}(z)\right] = -\frac{d}{ds}\varphi_{(0,1)}(0,z).$$

However, it follows from equation [2.38] that

$$\frac{d}{ds}\varphi_{(0,1)}(0,z) = \infty.$$

Therefore, we conclude that the cdf of $\tau_{(0,1)}(z)$ is a heavy tailed distribution.

REMARK 2.4.– *The distribution of the first collision of two particles in the Markov case, i.e. when* $G(t) = \left(1 - e^{-\lambda t}\right)\mathbf{1}_{\{t\geq 0\}}$, *was studied in Pogorui (2012b). It was proved that*

$$\varphi_{(0,1)}(s,z) = e^{-\frac{z}{2v}\sqrt{s^2+4\lambda s}}$$

and

$$\varphi_{(1,0)}(s,z) = \frac{s + 2\lambda - \sqrt{s^2+4\lambda s}}{s + 2\lambda + \sqrt{s^2+4\lambda s}}\, e^{-\frac{z}{2v}\sqrt{s^2+4\lambda s}}.$$

2.2.3. Distribution of the first collision of two particles with finite expectation

Now we are interested in distributions of $G(x)$ such that there exist the expectations $\mathbf{E}\left[\tau_{(0,1)}(z)\right] < \infty$ and $\mathbf{E}\left[\tau_{(1,0)}(z)\right] < \infty$.

Taking into account that

$$m_{(k_1,k_2)}(z) = \mathbf{E}\left[\tau_{(k_1,k_2)}(z)\right] = -\frac{\partial\varphi_{(k_1,k_2)}(0,z)}{\partial s},$$

it follows from equations [2.30] to [2.31] that

$$m_{(0,1)}(z) = \frac{z}{2v}\left(1 - G\left(\frac{z}{2v}\right)\right)^2 + \frac{z}{v^2}\int_0^\infty M(x)\,dx \int_0^z M\left(\frac{z-u}{2v}\right) du$$

$$+\frac{1}{v}\int_0^\infty M(x)\,x\,dx \int_0^z M\left(\frac{z-u}{2v}\right) du$$

$$-\frac{1}{v^2}\int_0^\infty M(x)\,dx \int_0^z M\left(\frac{z-u}{2v}\right) u\,du$$

$$+\frac{1}{2v}\int_0^\infty M(x)\,dx \int_0^z M\left(\frac{z-u}{2v}\right)\left(m_{(0,1)}(u) + m_{(1,0)}(u)\right) du,$$

$$m_{(1,0)}(z) = -\frac{z}{v^2}\int_0^\infty M(x)\,dx \int_z^\infty M\left(\frac{u-z}{2v}\right) du$$

$$+\frac{1}{v}\int_0^\infty M(x)\,x\,dx \int_z^\infty M\left(\frac{u-z}{2v}\right) du$$

$$+\frac{1}{v^2}\int_0^\infty M(x)\,dx \int_z^\infty M\left(\frac{u-z}{2v}\right) u\,du$$

$$+\frac{1}{2v}\int_0^\infty M(x)\,dx \int_z^\infty M\left(\frac{u-z}{2v}\right)\left(m_{(0,1)}(u) + m_{(1,0)}(u)\right) du.$$

Therefore, adding the last two equations, we have

$$m_{(0,1)}(z) + m_{(1,0)}(z) = F(z)$$

$$+\frac{1}{2v}\int_0^\infty M(x)\,dx \int_0^\infty M\left(\frac{|z-u|}{2v}\right)\left(m_{(0,1)}(u) + m_{(1,0)}(u)\right) du, \quad [2.39]$$

where $F(z)$ is the sum of the terms on the right sides of these equations, which do not contain $m_{(0,1)}(u) + m_{(1,0)}(u)$.

Let us show that there exists a cdf $G(x)$ such that $g(x)\,(1 - G(x)) = \frac{1}{\pi}K_0(x)$, $x \geq 0$, where $K_0(x)$ is the modified Bessel function of the second kind, namely

$$K_0(x) = \int_0^\infty \frac{\cos \omega x}{\sqrt{1+\omega^2}}\, d\omega.$$

It is well known that $\int_0^\infty K_0(x)dx = \frac{\pi}{2}$.

Consider the following equation:

$$\frac{dy(x)}{dx}\,(1 - y(x)) = \frac{1}{\pi}K_0(x). \quad [2.40]$$

Taking into account that $y(0) = 0$ and $\lim_{x\to+\infty} y(x) = 1$, it follows from equation [2.40] that

$$y(x) = 1 - \sqrt{1 - \frac{2}{\pi}\int_0^x K_0(t)\,dt}.$$

Therefore, for the distribution $G(x) = 1 - \sqrt{1 - \frac{2}{\pi}\int_0^x K_0(t)\,dt}$ and $v = 1/2$, equation [2.39] has the following form, for $z > 0$,

$$m_{(0,1)}(z)+m_{(1,0)}(z)-\frac{1}{2\pi}\int_0^\infty K_0\left(|z-u|\right)\left(m_{(0,1)}(u)+m_{(1,0)}(u)\right)du \quad [2.41]$$

$$= F(z).$$

Equation [2.41] is a particular case of the Grinberg–Fok integral equation, which is used in the theory of boundary refraction of electromagnetic waves (Krein 1962; Gakhov and Cherskii 1978), where it was shown that equation [2.41] has a unique solution with respect to $m_{(0,1)}(z) + m_{(1,0)}(z)$. Hence, $m_{(0,1)}(z)$ and $m_{(1,0)}(z)$ are finite for all $z > 0$, and as a result $m_{(0,0)}(z)$ and $m_{(1,1)}(z)$ are also finite.

PART 2

Financial Applications

3

Asymptotic Estimation for Application of the Telegraph Process as an Alternative to the Diffusion Process in the Black–Scholes Formula

In this chapter, we study the one-dimensional transport process in the case of disbalance. In the hydrodynamic limit, this process approximates the diffusion process on the line. The application of the telegraph process for option prices was studied in: Ratanov (2007, 2010); Ratanov and Melnikov (2008); López and Ratanov (2012); and Kolesnik and Ratanov (2013). Here, we give an application of the asymptotic result from Pogorui and Rodríguez-Dagnino (2009a) to derive an analog of the Black–Scholes formula.

3.1. Asymptotic expansion for the singularly perturbed random evolution in Markov media in the case of disbalance

Let $\{\xi(t), t \geq 0\}$ be a Markov process on the phase space $\{0, 1\}$ with the infinitesimal matrix

$$Q = \lambda \begin{pmatrix} -1 & 1 \\ 1 & -1 \end{pmatrix}.$$

Now, consider the random velocity model or transport process

$$x(t) = x_0 + \int_0^t v(s)ds, \qquad [3.1]$$

where

$$v(t) = \begin{cases} v_0, & \text{when } \xi(t) = 0; \\ v_1, & \text{when } \xi(t) = 1. \end{cases}$$

The infinitesimal operator A of the bivariate process

$$\{\varsigma(t) = (x(t), \xi(t)), \ t \geq 0\}$$

is of the following form:

$$A\varphi(x,0) = v_0 \frac{\partial}{\partial x}\varphi(x,0) + \lambda\varphi(x,1) - \lambda\varphi(x,0)$$

$$A\varphi(x,1) = v_1 \frac{\partial}{\partial x}\varphi(x,1) + \lambda\varphi(x,0) - \lambda\varphi(x,1), \qquad [3.2]$$

where $\varphi \in D(A)$ is the domain of the operator A, and $x \in \mathbb{R}$.

We can interpret this operator in the following equivalent manner: Assume $\mathbb{Z} = \mathbb{R} \times \{0,1\}$ and

$$T_t\varphi(x,i) = \int_{\mathbb{R}} \varphi(z) P\{\varsigma(t) \in dz \mid \varsigma(0) = (x,i)\}, \quad i \in \{0,1\}.$$

Then,

$$A\varphi(x,i) = \lim_{\triangle t \to 0+} \frac{T_{\triangle t}\varphi(x,i) - \varphi(x,i)}{\Delta t}.$$

Let us define $u_i(x,t) = T_t\varphi(x,i)$. Thus,

$$\frac{\partial}{\partial t} u_i(x,t) = AT_t\varphi(x,i) = Au_i(x,t).$$

Now, let us consider the scaled evolution equation

$$x_\varepsilon(t) = x_0 + \frac{1}{\varepsilon}\int_0^t v\left(\frac{s}{\varepsilon^2}\right) ds,$$

with corresponding velocities v_i/ε, $i \in \{0,1\}$. Since the Markov process $v\left(\frac{t}{\varepsilon^2}\right)$ has scaled time, then its infinitesimal operator is of the form $\frac{1}{\varepsilon^2}Q$.

Hence, we obtain the following system of Kolmogorov backward differential equations:

$$\frac{\partial}{\partial t} u_0^\varepsilon(x,t) = \frac{v_0}{\varepsilon}\frac{\partial}{\partial x} u_0^\varepsilon(x,t) + \frac{\lambda}{\varepsilon^2} u_1^\varepsilon(x,t) - \frac{\lambda}{\varepsilon^2} u_0^\varepsilon(x,t)$$

$$\frac{\partial}{\partial t} u_1^\varepsilon(x,t) = \frac{v_1}{\varepsilon}\frac{\partial}{\partial x} u_1^\varepsilon(x,t) + \frac{\lambda}{\varepsilon^2} u_0^\varepsilon(x,t) - \frac{\lambda}{\varepsilon^2} u_1^\varepsilon(x,t). \qquad [3.3]$$

Equations [3.3] can be written in the matrix form as follows:

$$\frac{\partial}{\partial t}\mathbf{u}^{\varepsilon}(x,t) = \frac{1}{\varepsilon}V\nabla\mathbf{u}^{\varepsilon}(x,t) + \frac{1}{\varepsilon^2}Q\mathbf{u}^{\varepsilon}(x,t) \qquad [3.4]$$

where

$$\mathbf{u}^{\varepsilon}(x,t) = \begin{pmatrix} u_0^{\varepsilon}(x,t) \\ u_1^{\varepsilon}(x,t) \end{pmatrix}, \qquad V = \begin{pmatrix} v_0 & 0 \\ 0 & v_1 \end{pmatrix},$$

and

$$V\nabla = \begin{pmatrix} v_0\frac{\partial}{\partial x} & 0 \\ 0 & v_1\frac{\partial}{\partial x} \end{pmatrix}, \qquad Q = \lambda\begin{pmatrix} -1 & 1 \\ 1 & -1 \end{pmatrix}.$$

Let us define

$$R_0 = \int_0^{\infty} \left(\Pi - e^{Qt}\right) dt,$$

where $e^{Qt} = \{p_{ij}(t);\ i,j \in \{0,1\}\}$ is the set of time-dependent transition probabilities, and

$$\Pi = \begin{pmatrix} \frac{1}{2} & \frac{1}{2} \\ \frac{1}{2} & \frac{1}{2} \end{pmatrix}$$

is the projector operator on $N(Q) = ker(Q)$ (Volume 1, Chapter 2, section 2.2).

Thus, $R_0 = \Pi - (Q+\Pi)^{-1}$ is the potential operator of $\xi(t)$ (Volume 1, Chapter 2, section 3.3) (Korolyuk and Turbin 1976).

The balance condition states that $\frac{v_0+v_1}{2} = 0$, i.e., $v_1 = -v_0$. This balance condition can also be expressed as $\Pi V \Pi = 0$ (see Volume 1, condition B5, theorem 4.1).

Now, we will consider the following disbalance condition: $v_0 = v + \Delta_1$ and $v_1 = -v - \Delta_2$, where $\Delta_i = \varepsilon a_i$, $i = 1, 2$, $\varepsilon > 0$. Then, it is easily verified that the infinitesimal operator of the process $\zeta_\varepsilon(t) = \left(x_\varepsilon(t),\ \xi\left(\frac{t}{\varepsilon^2}\right)\right)$ is of the following form:

$$A_\varepsilon = \frac{1}{\varepsilon}V\nabla + A\nabla + \frac{1}{\varepsilon^2}Q,$$

where

$$A = \begin{pmatrix} a_1 & 0 \\ 0 & a_2 \end{pmatrix}.$$

Denote as $u_i^\varepsilon(x,t) = T_t^\varepsilon \varphi(x,i) = \int_Z \varphi(z) P\{\zeta_\varepsilon(t) \in dz/\zeta_\varepsilon(0) = (x,i)\}$.

Then, similarly as above, we can write the matrix equation

$$\frac{\partial}{\partial t}\mathbf{u}^\varepsilon(x,t) = \frac{1}{\varepsilon}V\nabla\mathbf{u}^\varepsilon(x,t) + A\nabla\mathbf{u}^\varepsilon(x,t) + \frac{1}{\varepsilon^2}Q\mathbf{u}^\varepsilon(x,t). \qquad [3.5]$$

By using the technique of asymptotic expansion, as in the proof of Volume 1, theorem 4.1, we will find a solution of equation [3.5] in the following form:

$$\mathbf{u}^\varepsilon(x,t) = \mathbf{u}^{(0)}(x,t) + \sum_{n=1}^{\infty} \varepsilon^n \left(\mathbf{u}^{(n)}(x,t) + \mathbf{v}^{(n)}(x,t/\varepsilon^2)\right), \qquad [3.6]$$

where $\mathbf{u}^{(n)}(x,t)$, $n = 0,1,2,\ldots$ are the regular terms of the expansion, whereas $\mathbf{v}^{(n)}(x,t/\varepsilon^2)$, $n = 1,2,\ldots$ are the singular ones.

Then, by substituting equation [3.6] into [3.5], we obtain

$$\begin{aligned}
&Q\mathbf{u}^{(0)}(x,t) = 0, \qquad Q\mathbf{u}^{(1)}(x,t) + V\nabla\mathbf{u}^{(0)}(x,t) = 0,\\
&Q\mathbf{u}^{(2)}(x,t) + V\nabla\mathbf{u}^{(1)}(x,t) + A\nabla\mathbf{u}^{(0)}(x,t) - \frac{\partial}{\partial t}\mathbf{u}^{(0)}(x,t) = 0, \qquad [3.7]\\
&Q\mathbf{u}^{(k+2)}(x,t) + V\nabla\mathbf{u}^{(k+1)}(x,t) + A\nabla\mathbf{u}^{(k)}(x,t) - \frac{\partial}{\partial t}\mathbf{u}^{(k)}(x,t) = 0,
\end{aligned}$$

for $k \geq 0$.

Hence, $\mathbf{u}^{(0)}(x,t) \in N(Q)$, i.e, $\Pi\mathbf{u}^{(0)}(x,t) = \mathbf{u}^{(0)}(x,t)$.

It follows from equation [3.7] that

$$\mathbf{u}^{(1)}(x,t) = R_0 V\nabla\mathbf{u}^{(0)}(x,t) + c_1(t), \qquad [3.8]$$

where $c_1(t) \in N(Q)$.

Similarly,

$$\begin{aligned}
Q\mathbf{u}^{(2)}(x,t) &= \frac{\partial}{\partial t}\mathbf{u}^{(0)}(x,t) - V\nabla\mathbf{u}^{(1)}(x,t)\\
&= \frac{\partial}{\partial t}\mathbf{u}^{(0)}(x,t) - V\nabla R_0 V\nabla\mathbf{u}^{(0)}(x,t) - V\nabla c_1(t) - A\nabla\mathbf{u}^{(0)}(x,t). \qquad [3.9]
\end{aligned}$$

Multiplying equation [3.9] by the operator Π, we obtain

$$\Pi Q\mathbf{u}^{(2)}(x,t) = 0$$

$$= \frac{\partial}{\partial t}\mathbf{u}^{(0)}(x,t) - \Pi V\nabla R_0 V\nabla\Pi\mathbf{u}^{(0)}(x,t) - \Pi A\nabla\Pi\mathbf{u}^{(0)}(x,t). \qquad [3.10]$$

Here, we use the fact that $\Pi V\nabla c_1(t) = \Pi V\nabla\Pi c_1(t) = 0$.

Therefore, the main term of the expansion [3.6] satisfies the diffusion equation [3.10].

We can obtain similar equations for the rest of the terms, but we are only interested in the main term $\mathbf{u}^{(0)}(x,t)$.

Now, let us write the matrix equation [3.5] in the following form:

$$\begin{pmatrix} \frac{\partial}{\partial t} - \frac{v}{\varepsilon}\frac{\partial}{\partial x} - a_1\frac{\partial}{\partial x} + \frac{\lambda}{\varepsilon^2} & \frac{\lambda}{\varepsilon^2} \\ \frac{\lambda}{\varepsilon^2} & \frac{\partial}{\partial t} + \frac{v}{\varepsilon}\frac{\partial}{\partial x} + a_2\frac{\partial}{\partial x} + \frac{\lambda}{\varepsilon^2} \end{pmatrix} \mathbf{u}^{(\varepsilon)}(x,t)$$

$$= \Psi\mathbf{u}^{(\varepsilon)}(x,t) = 0.$$

It is easy to verify that the function $f_\varepsilon(x,t) = u_0^\varepsilon(x,t) + u_1^\varepsilon(x,t)$ is the solution of the following equation:

$$\det(\Psi)\, f_\varepsilon(x,t) = 0. \qquad [3.11]$$

Writing equation [3.11] in more detail, we have

$$\left[\left(2\lambda\frac{\partial}{\partial t} - v^2\frac{\partial^2}{\partial x^2} + \lambda(a_2 - a_1)\frac{\partial}{\partial x}\right)\right.$$
$$\left. + \varepsilon^2\left(\frac{\partial^2}{\partial t^2} + (a_2 - a_1)\frac{\partial^2}{\partial x\partial t} - \frac{v(a_2 + a_1)}{\varepsilon}\frac{\partial^2}{\partial x^2} - a_1a_2\frac{\partial^2}{\partial x^2}\right)\right] \qquad [3.12]$$
$$\times f_\varepsilon(x,t) = 0.$$

Let us define the notation $u^{(0)}(x,t) = (u_0(x,t), u_1(x,t))$ and $f_0(x,t) = u_0(x,t) + u_1(x,t)$. Since $\mathbf{u}^\varepsilon(x,t) > \mathbf{u}^{(0)}(x,t)$ as $\varepsilon > 0$ (Korolyuk and Korolyuk 1999), we have $\lim_{\varepsilon\to 0} f_\varepsilon(x,t) = f_0(x,t)$.

Hence, it follows from equation [3.12] that

$$\left(\frac{\partial}{\partial t} - \frac{v^2}{2\lambda}\frac{\partial^2}{\partial x^2} + \frac{a_2 - a_1}{2}\frac{\partial}{\partial x}\right) f_0(x,t) = 0. \qquad [3.13]$$

So, if $a_2 \neq a_1$, then $f_0(x,t)$ satisfies the diffusion equation [3.13] with drift coefficient $\frac{a_1-a_2}{2}$ and diffusion coefficient $\frac{v^2}{2\lambda}$.

3.2. Application: Black–Scholes formula

The well-known geometric Brownian motion (GBM) is used to model a price S_t of a stock (Shreve 2004) at time t such that

$$S_t = S_0 \exp(\mu t + \sigma W_t), \qquad [3.14]$$

where μ is the drift, σ is the volatility of the stock and $\{W_t\}$ is the Wiener process.

By substituting $\mu = \frac{a_1 - a_2}{2}$ and $\sigma = \frac{v}{\sqrt{2\lambda}}$ in equation [3.11], we obtain the diffusion equation for the process $\{\mu t + \sigma W_t\}$. So, we can obtain the Black–Scholes formula by considering an exponential Brownian motion for the share price S_t.

The Wiener process provides a mathematically consistent model for Brownian motion. However, it has the following drawbacks to capture the physics of many applications:

1) modulus of the velocity is almost always infinite at any instant of time,

2) zero length of free path,

3) the path function of a particle is almost surely nowhere differentiable, and its Hausdorff dimension is equal to 1.5, i.e. the path function is fractal.

However, the actual movement of a physical particle, as well as the actual evolution of share prices S_t, are barely justified as fractal quantities. Taking into account these considerations, we propose using the process given by equation [3.12] instead of the diffusion process given in equation [3.13] for most applications. This recommendation is based on the fact that this process does not suffer from the drawback mentioned above. For instance, the process in equation [3.12] has almost always a finite modulus of velocity $V_\varepsilon = \frac{v}{\varepsilon}$, non-zero free paths that depend on $\Lambda_\varepsilon = \frac{\lambda}{\varepsilon^2}$ and its trajectories are almost surely differentiable.

Some difficulties may arise for calculating the coefficients V_ε and Λ_ε. For such a purpose, we suggest considering the time between two consecutive renewal epochs (or equivalently two consequent impacts of a particle or two consequent changes of price) as an exponential distributed random variable with parameter Λ_ε. Then, Λ_ε can be estimated from experimental data as well as the velocity V_ε.

Hence, instead of the GBM we propose the following formula for the price S_t of a stock at time t:

$$S_t^\epsilon = S_0 \exp\left(\varepsilon^{-1} \int_0^t v\left(\frac{s}{\varepsilon^2}\right) ds\right),$$

where $v(s)$ is defined in [3.1].

This model can be used in the modeling of cash flow in high-frequency and algorithmic trading. As long as transactions happen every millisecond, then $v(t/\epsilon^2)$ is switched very quickly between two states $\{v_0, v_1\}$ according to the Markov chain $\xi(t/\epsilon^2)$, because of the scale $1/\epsilon^2$, meaning that S_t^ϵ is changing very quickly as well. Instead of time t as milliseconds, we use a longer time interval t/ϵ^2 which is needed, for example, for liquidation, acquisition, market making, purposes, etc. In this way, we can apply our asymptotical results considered above. Below we show how to get an option pricing formula, namely the Black–Scholes formula, for our model of a stock price. Of course, this result may be applied to other problems in mathematical finance, such as portfolio optimization and optimal control.

Thus, applying the results from the previous subsection we can state that

$$S_t^\epsilon \to_{\epsilon\to 0} \hat{S}_t,$$

weakly in Skorokhod topology, where

$$\hat{S}_t = S_0 \exp\left\{\frac{a_1 - a_2}{2}t + \frac{v^2}{2\lambda}W_t\right\}.$$

According to Itô's formula (Shreve 2004), $\hat{S}_t$ satisfies the following SDE:

$$\hat{S}_t = \hat{\mu}\hat{S}_t dt + \frac{v^2}{2\lambda}\hat{S}_t W_t,$$

where

$$\hat{\mu} := \left\{\frac{\frac{a_1-a_2}{2} + \frac{v^4}{8\lambda^2}}{\frac{v^2}{2\lambda}}\right\}.$$

Let us introduce the following process:

$$Z(t) := \exp\left\{\left(\frac{\hat{\mu} - r}{v^2/2\lambda}\right)W_t - \frac{1}{2}\left(\frac{\hat{\mu} - r}{v^2/2\lambda}\right)^2 t\right\},$$

where $r > 0$ is the interest rate.

Then $\mathbf{E}_P[Z(t)] = 1$, and this process is a positive martingale, according to Novikov's result (Shreve 2004). Define the following new probability measure Q (recall $\mathbf{E}_P[Z(t)] = 1$) on a complete probability space $(\Omega, \mathcal{F})$:

$$Q(A) := \mathbf{E}_P[Z(T)\mathbf{1}_A],$$

where $\mathbf{1}_A$ is an indicator of the set $A \in \mathcal{F}$, and $Z(t)$ is defined above.

Let us also define the following process:

$$\hat{W}_t := W_t - \left(\frac{\hat{\mu} - r}{v^2/2\lambda}\right)t,$$

where $\hat{\mu}$ is defined above.

According to Girsanov's theorem (Shreve 2004), under the probability measure Q, the stochastic process $\hat{W}_t$ is a standard Wiener process. We call measure Q a risk-neutral or martingale measure. Then under the risk-neutral measure Q our stock price $\hat{S}_t$ satisfies the following SDE:

$$d\hat{S}_t = r\hat{S}_t dt + \frac{v^2}{2\lambda}\hat{S}_t d\hat{W}_t.$$

Therefore, we can write the Black–Scholes formula for European call option price $C(t)$ for our model:

$$C(t) = S_0 N(d_1) - Ke^{-r(T-t)}N(d_2), \quad [3.15]$$

where

$$d_1 := \frac{\ln(S_0/K) + (r + v^4/8\lambda^2)(T-t)}{(v^2/2\lambda)\sqrt{T-t}},$$

$$d_2 := d_1 - (v^2/2\lambda)\sqrt{T-t}, \quad [3.16]$$

and

$$N(x) = \Phi(x) = \frac{1}{\sqrt{2\pi}}\int_{-\infty}^{x} e^{-\frac{u^2}{2}} du$$

is the cumulative distribution function of a zero mean normal random variable with unit variance, K is a strike price and T is the maturity.

NUMERICAL EXAMPLE.– Suppose that $S_0 = \$50, K = 50, r = 0.01, v = 0.01$, $\lambda = 0.01, T = 1$. Then, according to formulas [3.15] and [3.16], we have the following European call option price at time $t = 0$:

$$C(0) = 50 \times 0.9774 - 50 \times e^{-0.01\times 1} \times 0.9771 = \$0.4996$$

or

$$C(0) = 49.96 \text{ cents}.$$

Below we present some graphs of dynamics $C(t)$ dependent on λ (upon fixed v), v (upon fixed λ), and v and λ.

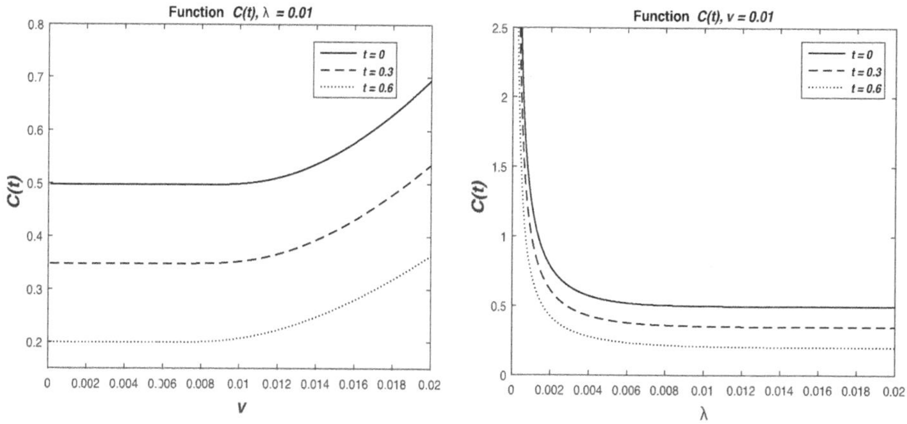

Figure 3.1. *Dependence of European call option price on v (left) and λ (right)*

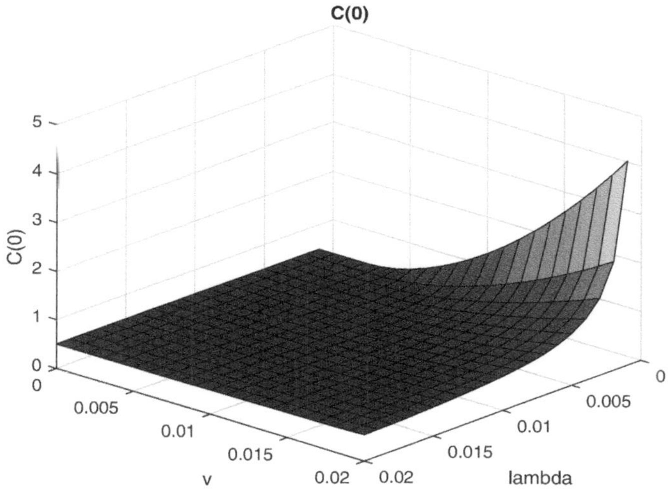

Figure 3.2. *Dependence of European call option price on v and λ . For a color version of this figure, see www.iste.co.uk/pogorui/random2.zip*

4

Variance, Volatility, Covariance and Correlation Swaps for Financial Markets with Markov-modulated Volatilities

In this chapter, we price variance, volatility, covariance and correlation swaps for financial markets with Markov-modulated volatilities. As an example, we consider stochastic volatility driven by a two-state continuous Markov chain. In this case, some numerical examples are presented for VIX and VXN volatility indeces (S&P 500 and NASDAQ-100, respectively, since January 2004 to June 2012). We also use VIX (January 2004 to June 2012) to price variance and volatility swaps for the two-state Markov-modulated volatility and to present a numerical result in this case. The results of this chapter are mainly based on a paper by Salvi and Swishchuk (2012a).

4.1. Volatility derivatives

4.1.1. *Types of volatilities*

Volatility σ is the standard deviation of the change in value of a financial instrument with a specific horizon. It is often used to quantify the risk of the instrument over that time period. The higher the volatility, the riskier the security.

Historical volatility is the volatility of a financial instrument based on historical returns. It is a standard deviation that uses historical (daily, weekly, monthly, quarterly, yearly) price data to empirically measure the volatility of a market or instrument in the past.

The annualized volatility σ is the standard deviation of the instrument's logarithmic returns in a year:

$$\sigma := \sqrt{\frac{1}{n-1}\sum_{i=1}^{n}(R_i - \bar{R})^2},$$

where $R_i := \ln\left(\frac{S_{t_i}}{S_{t_{i-1}}}\right)$, $\bar{R} := \frac{1}{n}\sum_{i=1}^{n}\ln\left(\frac{S_{t_i}}{S_{t_{i-1}}}\right)$, S_{t_i} is an asset price at time t_i, $i = 1, 2, \dots, n$.

Implied volatility is related to historical volatility, however the two are distinct. Historical volatility is a direct measure of the movement of the underlier's price (realized volatility) over recent history. Implied volatility, in contrast, is set by the market price of the derivative contract itself, and not the underlier. Therefore, different derivative contracts on the same underlier have different implied volatilities. Most derivative markets exhibit persistent patterns of volatilities varying by strike. The pattern displays different characteristics for different markets. In some markets, the patterns form a smile curve. In others, such as equity index options markets, they form a skewed curve. This is what inspired the name "volatility skew". For markets where the graph is sloping downwards, such as equity options, the term *"volatility skew"* is often used. For other markets, such as FX options or equity index options, where the typical graph turns up at either end, the more familiar term *"volatility smile"* is used. In practice, either the term "volatility smile" or "volatility skew" may be used to refer to the general phenomenon of volatilities varying by strike.

The models by Black and Scholes (1973), continuous-time (B,S)-security market, and (Cox and Ross 1976; Cox *et al.* 1979), discrete-time (B,S)-security market (binomial tree), are unable to explain the negative skewness and leptokurticity (fat tail) commonly observed in the stock markets. The famous implied-volatility smile would not exist under their assumptions.

Given the prices of call or put options across all strikes and maturities, we may deduce the volatility that produces those prices via the full Black–Scholes equation (Black and Scholes 1973; Dupire 1994; Derman and Kani 1994). This function has come to be known as *local volatility*. The local volatility-function of the spot price S_t and time $t : \sigma \equiv \sigma(S_t, t)$; see (Dupire 1994) formulae for local volatility.

Level-dependent volatility (e.g. CEV or firm model) - function of the spot price alone: to have a smile across strike price, we need σ to depend on $S : \sigma \equiv \sigma(S_t)$. In this case, the volatility and stock price changes are now highly correlated.

4.1.2. *Models for volatilities*

In the early 1970s, Black and Scholes (1973) made a major breakthrough by deriving pricing formulas for vanilla options written on the stock. The Black–Scholes model assumes that the volatility term is a constant. This assumption is not always satisfied by real-life options, as the probability distribution of an equity has a fatter left tail and a thinner right tail than the lognormal distribution (see Hull 2000), and the assumption of constant volatility σ in financial models (such as the original Black–Scholes model) is incompatible with derivative prices observed in the market. Stochastic volatility models are used in the field of quantitative finance to evaluate derivative securities, such as options and swaps. By assuming that the volatility of the underlying price is a stochastic process rather than a constant, it becomes possible to model derivatives more accurately.

The above issues have been addressed and studied in several ways:

1) volatility is assumed to be a deterministic function of time: $\sigma \equiv \sigma(t)$ (see Wilmott *et al.* 1995; Merton 1973) extended the term structure of volatility to $\sigma := \sigma_t$ (deterministic function of time), with the implied volatility for an option of maturity T, given by $\hat{\sigma}_T^2 = \frac{1}{T}\int_0^T \sigma_u^2 du$;

2) volatility is assumed to be a function of time and the current level of the stock price $S(t)$: $\sigma \equiv \sigma(t, S(t))$ (see Hull 2000); the dynamics of the stock price satisfies the following stochastic differential equation:

$$dS(t) = \mu S(t)dt + \sigma(t, S(t))S(t)dW_1(t),$$

where $W_1(t)$ is a standard Wiener process;

3) the time variation of volatility involves an additional source of randomness, besides $W_1(t)$, represented by $W_2(t)$, and it is given by

$$d\sigma(t) = a(t, \sigma(t))dt + b(t, \sigma(t))dW_2(t),$$

where $W_2(t)$ and $W_1(t)$ (the initial Wiener process that governs the price process) may be correlated (see Buff 2002; Hull and White 1987; Heston 1993);

4) the volatility depends on a random parameter x such as $\sigma(t) \equiv \sigma(x(t))$, where $x(t)$ is a random process (see Elliott and Swishchuk 2007; Griego and Swishchuk 2000; Swishchuk 2000). Cox and Ross (1976) valued options for alternative stochastic processes. Harrison Pliska (1981) introduced and studied arbitrage and completeness of the Brownian market. Föllmerand Sondermann (1986) introduced and studied locally minimizing risk strategies. Hamilton (1989) introduced Markov switching into the econometric mainstream. Föllmer and Schweizer (1991) studied hedging under incomplete information, using the minimal martingale measure. Di Masi *et al.* (1994) obtained an option pricing formula for stochastic volatility, driven by a Markov chain in continuous time. Swishchuk (1997) obtained an option

pricing formula for a model with stochastic volatility, driven by a semi-Markov process. Gray (1996) combined GARCH effects with Markov switching. Griego and Swishchuk (2000) obtained the Black–Scholes formula for a market in a Markov random environment;

5) another approach is connected with stochastic volatility, namely, the uncertain volatility scenario (see Buff 2002). This approach is based on the uncertain volatility model developed in Avellaneda *et al.* (1995), where a concrete volatility surface is selected among a candidate set of volatility surfaces. This approach addresses the sensitivity question, by computing an upper bound for the value of the portfolio under arbitrary candidate volatility, and this is achieved by choosing the local volatility $\sigma(t, S(t))$ among two extreme values $\sigma_{\min}$ and $\sigma_{\max}$, such that the value of the portfolio is maximized locally;

6) the volatility $\sigma(t, S_t)$ depends on $S_t := S(t + \theta)$ for $\theta \in [-\tau, 0]$, namely, stochastic volatility with delay (see Kazmerchuk *et al.* 2002).

In approach (1), the volatility coefficient is independent of the current level of the underlying stochastic process $S(t)$. This is a deterministic volatility model, and the special case where σ is a constant, reduces to the well-known Black–Scholes model, which suggests changes in stock prices are lognormal distributed. But the empirical test by Bollerslev (1986) seems to indicate otherwise. One explanation for this problem of a lognormal model is the possibility that the variance of $\log(S(t)/S(t-1))$ changes randomly. This motivated (Chesney and Scott 1989), where the prices were analyzed for European options using the modified Black–Scholes model of foreign currency options and a random variance model. In their works, the results of Hull and White (1987), Scott (1987) and Wiggins (1987) were used in order to incorporate randomly changing variance rates.

In approach (2), several methods have been developed to derive the corresponding Black–Scholes formula: one can obtain the formula by using stochastic calculus and, in particular, the Itô's formula (see Shiryaev 1999).

A generalized volatility coefficient of the form $\sigma(t, S(t))$ is said to be *level dependent*. Because volatility and asset price are highly correlated, we only have one source of randomness given by $W_1(t)$. A time and level-dependent volatility coefficient makes the arithmetic more challenging and usually precludes the existence of a closed-form solution. However, the *arbitrage argument*, based on portfolio replication and a completeness of the market, remains unchanged.

The situation becomes different if the volatility is influenced by a second "non-tradable" source of randomness. This is addressed in approaches (3)–(5), where we usually obtain a *stochastic volatility model*, which is generally enough to include the deterministic model as a special case. The concept of stochastic volatility was introduced by Hull and White (1987), and subsequent development includes the work of Wiggins (1987); Johnson and Shanno (1987); Scott (1987); Stein and Stein (1991) and (Heston 1993). We also refer to Frey (1997) for an excellent survey on level-dependent and stochastic volatility models (see also Derman 1999).

Hobson and Rogers (1998) suggested a new class of non-constant volatility models, which can be extended to include the aforementioned level-dependent model and share many characteristics with the stochastic volatility model. The volatility is non-constant and can be regarded as an endogenous factor in the sense that it is defined in terms of the *past behavior* of the stock price. This is done in such a way that the price and volatility form a multidimensional Markov process.

The generalized auto-regression conditional heteroskedacity (GARCH) model (see Bollerslev 1986) is another popular model for estimating stochastic volatility. It assumes that the randomness of the variance process varies with the variance; this is opposed to the square root of the variance, as is assumed in the Heston model. The standard GARCH(1,1) model has the following form for the variance differential:

$$d\sigma_t = \kappa(\theta - \sigma_t)dt + \gamma\sigma_t dB_t.$$

The GARCH model has been extended via numerous variants, including the NGARCH, LGARCH, EGARCH and GJR-GARCH (see, e.g. Duan 1995).

4.1.3. *Variance and volatility swaps*

Volatility swaps are forward contracts on future realized stock volatility, whereas variance swaps are a similar contract on variance, the square of the future volatility; both of these instruments provide an easy way for investors to gain exposure to the future level of volatility.

A stock's volatility is the simplest measure of its risk or uncertainty. Formally, the volatility σ_R is the annualized standard deviation of the stock's returns during the period of interest, where the subscript R denotes the observed or "realized" volatility.

The easy way to trade volatility is to use volatility swaps, sometimes called realized volatility forward contracts, because they provide pure exposure to volatility (and only to volatility).

Demeterfi *et al.* (1999) explained the properties and the theory of both variance and volatility swaps. They derived an analytical formula for theoretical fair value in the presence of realistic volatility skews, and pointed out that volatility swaps can be replicated by dynamically trading the more straightforward variance swap.

Javaheri *et al.* (2002) discussed the valuation and hedging of a GARCH(1,1) stochastic volatility model. They used a general and flexible PDE approach to determine the first two moments of the realized variance in a continuous or discrete context. Then, they approximated the expected realized volatility via a convexity adjustment.

Brockhaus and Long (2000) provided an analytical approximation for the valuation of volatility swaps and analyzed other options with volatility exposure.

The working paper by Théoret *et al.* (2002) presented an analytical solution for the pricing of volatility swaps, proposed by Javaheri *et al.* (2002). They priced the volatility swaps within the framework of the GARCH(1,1) stochastic volatility model and applied the analytical solution to price a swap on volatility of the $S\&P60$ Canada Index (five-year historical period: 1997–2002).

Elliott and Swishchuk (2007) studied option pricing formulae and pricing swaps for a Markov-modulated Brownian process with jumps. Variance and volatility swaps for financial markets with stochastic volatility, which follow the Heston model, have been studied in Swishchuk (2004), and variance swaps for financial markets with stochastic volatilities with delay, have been studied in Swishchuk (2005).

4.1.4. *Covariance and correlation swaps*

Among the recent and new financial products are covariance and correlation swaps, which are useful for volatility hedging and speculation, using two different financial underlying assets.

For example, options dependent on exchange rate movements, such as those paying in a currency different from the underlying currency, have an exposure to movements of the correlation between the asset and the exchange rate; this risk may be eliminated by using a covariance swap.

A *covariance swap* is a covariance forward contract of the underlying rates S^1 and S^2, whose payoff at expiration is equal to

$$N(Cov_R(S^1, S^2) - K_{cov}),$$

where K_{cov} is a strike price, N is the notional amount and $Cov_R(S^1, S^2)$ is a covariance between the two assets S^1 and S^2.

A *correlation swap* is a correlation forward contract of two underlying rates S^1 and S^2 whose payoff at expiration is equal to:

$$N(Corr_R(S^1, S^2) - K_{corr}),$$

where $Corr(S^1, S^2)$ is a realized correlation of the two underlying assets S^1 and S^2, K_{corr} is a strike price and N is the notional amount.

Pricing covariance swaps, from a theoretical point of view, are similar to pricing variance swaps, since

$$Cov_R(S^1, S^2) = \frac{1}{4}\{\sigma_R^2(S^1 S^2) - \sigma_R^2(S^1/S^2)\}$$

where S^1 and S^2 are two given assets, $\sigma_R^2(S)$ is a variance swap for underlying assets and $Cov_R(S^1, S^2)$ is a realized covariance of the two underlying assets S^1 and S^2.

Thus, we need to know variances for $S^1 S^2$ and for S^1/S^2 (see section 4.2 for details).

Correlation $Corr_R(S^1, S^2)$ is defined as follows:

$$Corr_R(S^1, S^2) = \frac{Cov_R(S^1, S^2)}{\sqrt{\sigma_R^2(S^1)}\sqrt{\sigma_R^2(S^2)}},$$

where $Cov_R(S^1, S^2)$ is defined above and $\sigma_R^2(S^1)$ in section 4.3.1 of Volume 1.

Given two assets S_t^1 and S_t^2 with $t \in [0, T]$, sampled on days $t_0 = 0 < t_1 < t_2 < \dots < t_n = T$ between today and maturity T, the log-return of each asset is $R_i^j := \log\left(\frac{S_{t_i}^j}{S_{t_{i-1}}^j}\right)$, $i = 1, 2, \dots, n, \quad j = 1, 2.$

Covariance and correlation can be approximated by

$$Cov_n(S^1, S^2) = \frac{n}{(n-1)T}\sum_{i=1}^{n} R_i^1 R_i^2$$

and

$$Corr_n(S^1, S^2) = \frac{Cov_n(S^1, S^2)}{\sqrt{Var_n(S^1)}\sqrt{Var_n(S^2)}},$$

respectively.

4.1.5. *A brief literature review*

The literature devoted to the volatility derivatives is growing. Here, we give a short overview of the latest development in this area. The non-Gaussian Ornstein–Uhlenbeck stochastic volatility model was used by Benth *et al.* (2007) to study volatility and variance swaps. Broadie and Jain (2008b) evaluated price and hedging strategy for volatility derivatives in the Heston square root stochastic volatility model, and they compared results from various models, in order to investigate the effect of jumps and discrete sampling on variance and volatility swaps

(Broadie and Jain 2008a). Pure jump processes with independent increment return models were used by Carr *et al.* (2005) to price derivatives written on realized variance, and were subsequent developed by Carr and Lee (2007). We also refer to Carr and Lee (2009) for a good survey on volatility derivatives. Da Fonseca *et al.* (2009) analyzed the influence of variance and covariance swap in a market, by solving a portfolio optimization problem in a market with risky assets and volatility derivatives. Correlation swap price has been investigated by Bossu (2005, 2007) for components of an equity index using a statistical method. The paper by Drissien *et al.* (2009) discusses the price of correlation risk for equity options. Pricing volatility swaps under Heston's model with regime-switching, and pricing options under a generalized Markov-modulated jump-diffusion model, are discussed in Elliott *et al.* (2007b,a), respectively. The paper by Howison *et al.* (2004) considers the pricing of a range of volatility derivatives, including volatility and variance swaps and swaptions. The pricing options on realized variance in the Heston model, with jumps in returns and volatility, are studied in Sepp (2008). An analytical closed-form pricing of pseudo-variance, pseudo-volatility, pseudo-covariance and pseudo-correlation swaps is studied in Swishchuk *et al.* (2002). The paper by Windcliff *et al.* (2006) investigates the behavior and hedging of discretely observed volatility derivatives.

The remaining chapter is organized as follows. The martingale representation of a Markov processes is considered in section 4.2. Variance and volatility swaps for financial markets with Markov-modulated stochastic volatilities are studied in section 4.3. The pricing of covariance and correlation swaps for two risky assets for financial markets with Markov-modulated stochastic volatilities is presented in section 4.4. In section 4.5, we consider an example for variance, volatility, covariance and correlation swaps for Markov-modulated volatility, driven by a two-state continuous Markov chain. A numerical example for variance and volatility swaps pricing using the S&P500 VIX index (January 2004–June 2012) is given in section 4.6.1. A numerical example for covariance and correlation swaps using the S&P500 VIX and NASDAQ-100 VXN indeces (January 2004–June 2012) is presented in section 4.6.2. The Appendix contains first-order correction formula for correlation swaps.

4.2. Martingale representation of a Markov process

Let $(\Omega, \mathcal{F}, (\mathcal{F}_t)_{t\in\mathbb{R}_+}, \mathbb{P})$ be a filtered probability space, with a right-continuous filtration $(\mathcal{F}_t)_{t\in\mathbb{R}_+}$ and probability measure $\mathbb{P}$. Let $(X, \mathcal{X})$ be a measurable space and $(x_t)_{t\in\mathbb{R}_+}$ be a $(X, \mathcal{X})$-valued Markov process with generator Q. The following two results allow us to associate a martingale to this process, and to obtain its quadratic variation; we refer to Elliott and Swishchuk (2007) for the proofs.

PROPOSITION 4.1 (Elliott and Swishchuk 2007).– Let $(x_t)_{t\in\mathbb{R}_+}$ be a Markov process with generator Q and $f \in$ Domain (Q), then

$$m_t^f := f(x_t) - f(x_0) - \int_0^t Qf(x_s)ds \qquad [4.1]$$

is a zero-mean martingale with respect to $\mathcal{F}_t := \sigma\{y(s); 0 \leq s \leq t\}$.

Let us evaluate the quadratic variation of this martingale.

PROPOSITION 4.2 (Elliott and Swishchuk 2007).– Let $(x_t)_{t\in\mathbb{R}_+}$ be a Markov process with generator Q, $f \in$ Domain (Q) and $(m_t^f)_{t\in\mathbb{R}_+}$ as its associated martingale, then

$$\langle m^f \rangle_t := \int_0^t [Qf^2(x_s) - 2f(x_s)Qf(x_s)]ds \qquad [4.2]$$

is the quadratic variation of m^f.

In the following, it will be useful to consider the quadratic covariation of two martingales associated with a couple of generic functions of a Markov process.

PROPOSITION 4.3.– Let $(x_t)_{t\in\mathbb{R}_+}$ be a Markov process with generator Q, $f, g \in$ Domain (Q) such that $fg \in$ Domain (Q). Denote by $(m_t^f)_{t\in\mathbb{R}_+}, (m_t^g)_{t\in\mathbb{R}_+}$ their associated martingales. Then,

$$\begin{aligned}&\langle f(x.), g(x.)\rangle_t \\ &:= \int_0^t \{Q(f(x_s)g(x_s)) - [g(x_s)Qf(x_s) + f(x_s)Qg(x_s)]\}\, ds \qquad [4.3]\end{aligned}$$

is the quadratic covariation of f and g.

DEMONSTRATION.– First of all, we note that

$$\begin{aligned}m_t^f m_t^g &= f(x_t)g(x_t) - f(x_t)\int_0^t Qg(x_s)ds - g(x_t)\int_0^t Qf(x_s)ds \\ &+ \int_0^t Qf(x_s)ds \int_0^t Qg(x_s)ds \\ &= f(x_t)g(x_t) - m_t^f \int_0^t Qg(x_s)ds - m_t^g \int_0^t Qf(x_s)ds \\ &- \int_0^t Qf(x_s)ds \int_0^t Qg(x_u)du. \qquad [4.4]\end{aligned}$$

Moreover,

$$d\left[m_t^f \int_0^t Qg(x_s)ds + m_t^g \int_0^t Qf(x_s)ds + \int_0^t Qf(x_s)ds \int_0^t Qg(x_u)du\right]$$
$$= \left(\int_0^t Qg(x_s)ds\right) dm_t^f + \left(\int_0^t Qf(x_s)ds\right) dm_t^g$$
$$+ m_t^f Qg(x_t)dt + m_t^g Qf(x_t)dt$$
$$+ \left(\int_0^t Qg(x_s)ds\right) Qf(x_t)dt + \left(\int_0^t Qf(x_s)ds\right) Qg(x_t)dt. \qquad [4.5]$$

Now using the expression for m^f and m^g (see proposition 4.1), we obtain

$$d\left[m_t^f \int_0^t Qg(x_s)ds + m_t^g \int_0^t Qf(x_s)ds + \int_0^t Qf(x_s)ds \int_0^t Qg(x_u)du\right]$$
$$= \left(\int_0^t Qg(x_s)ds\right) dm_t^f + \left(\int_0^t Qf(x_s)ds\right) dm_t^g \qquad [4.6]$$
$$+ f(x_t)Qg(x_t)dt - g(x_t)Qf(x_t)dt.$$

By using [4.6], equation [4.4] becomes

$$m_t^f m_t^g = f(x_t)g(x_t)$$
$$-\left[\int_0^t \left(\int_0^s Qg(x_u)du\right) dm_s^f + \int_0^t \left(\int_0^s Qf(x_u)du\right) dm_s^g\right] \qquad [4.7]$$
$$-\int_0^t \left[f(x_s)Qg(x_s) + g(x_s)Qf(x_s)\right] ds.$$

Adding and subtracting $\int_0^t Q(f(x_s)g(x_s))ds$ on the right-hand side of the previous equation, we have

$$m_t^f m_t^g = f(x_t)g(x_t) - \int_0^t Q(f(x_s)g(x_s))ds$$
$$-\left[\int_0^t \left(\int_0^s Qg(x_u)du\right) dm_s^f + \int_0^t \left(\int_0^s Qf(x_u)du\right) dm_s^g\right] \qquad [4.8]$$
$$+\int_0^t \left[Q(f(x_s)g(x_s)) - f(x_s)Qg(x_s) - g(x_s)Qf(x_s)\right] ds.$$

Since $fg \in Domain\ (Q)$, then (see proposition 4.1), and for $t \in \mathbb{R}_+$

$$f(x_t)g(x_t) - \int_0^t Q(f(x_s)g(x_s))ds, \qquad [4.9]$$

is a martingale. The term in square brackets on the right-hand side of [4.8] is a martingale as well. Therefore, for $t \in \mathbb{R}_+$

$$m_t^f m_t^g - \int_0^t [Q(f(x_s)g(x_s)) - f(x_s)Qg(x_s) - g(x_s)Qf(x_s)]\, ds, \qquad [4.10]$$

is a martingale, and we have that

$$\langle f(x.), g(x.)\rangle_t$$
$$= \int_0^t [Q(f(x_s)g(x_s)) - f(x_s)Qg(x_s) - g(x_s)Qf(x_s)]\, ds. \qquad [4.11]$$

Now, we are able to evaluate the expectation of a generic function of a Markov process.

PROPOSITION 4.4 (Elliott and Swishchuk 2007).– Let $(x_t)_{t\in\mathbb{R}_+}$ be a Markov process with generator Q and $f \in$ Domain (Q), then

$$\mathbf{E}\{f(x_t)\} = e^{tQ} f(x_0). \qquad [4.12]$$

DEMONSTRATION.– From proposition 4.1, we have

$$\mathbf{E}\{f(x_t)\} = f(x_0) + \int_0^t Q\,\mathbf{E}\{f(x_s)\}ds, \qquad [4.13]$$

and after solving this differential equation we obtain the result.

PROPOSITION 4.5.– Let $(x_t)_{t\in\mathbb{R}_+}$ be a Markov process with generator Q, $f, g \in$ Domain (Q), such that $fg \in$ Domain (Q) and denote by $(m_t^f)_{t\in\mathbb{R}_+}, (m_t^g)_{t\in\mathbb{R}_+}$ their associated martingales. Then,

$$\mathbf{E}\{f(x_t)g(x_t)\} = e^{tQ} f(x_0)g(x_0). \qquad [4.14]$$

DEMONSTRATION.– Applying Itô's lemma to the product fg, we obtain

$$d(f(x_t)g(x_t)) = f(x_t)dg(x_t) + g(x_t)df(x_t) + d\langle f(x.), g(x.)\rangle_t, \qquad [4.15]$$

and using propositions 4.1 and 4.3 we have

$$d(f(x_t)g(x_t)) = Q\,[f(x_t)g(x_t)]\, dt + f(x_t)dm_t^g + g(x_t)dm_t^f. \qquad [4.16]$$

This stochastic differential equation can be written in integral form; taking the expectation of its integral representation, we have

$$\mathbf{E}\{f(x_t)g(x_t)\} = f(x_0)g(x_0) + \int_0^t Q\,[f(x_s)g(x_s)]\,ds. \qquad [4.17]$$

Solving this differential equation, we obtain the result

$$\mathbf{E}\{f(x_t)g(x_t)\} = e^{tQ} f(x_0)g(x_0). \qquad [4.18]$$

4.3. Variance and volatility swaps for financial markets with Markov-modulated stochastic volatilities

Let us consider a financial market with only two securities, the risk-free bond and the stock. Let us suppose that the stock price $(S_t)_{t\in\mathbb{R}_+}$ satisfies the following stochastic differential equation:

$$dS_t = S_t(r\,dt + \sigma(x_t)dw_t), \qquad [4.19]$$

where w is a standard Wiener process independent of the Markov process $(x_t)_t$. In this model, the volatility is stochastic, and it is interesting to study the property of σ and in particular, how to price future contracts on realized variance and volatility.

The following results concern the expectation of variance, and they are a simple application of propositions 4.1 and 4.4.

PROPOSITION 4.6.– Suppose that $\sigma \in Domain\,(Q)$. Then,

$$\mathbf{E}\{\sigma^2(x_t)|\,\mathcal{F}_u\} = \sigma^2(x_u) + \int_u^t Q\,\mathbf{E}\{\sigma^2(x_s)|\,\mathcal{F}_u\}ds \qquad [4.20]$$

for any $0 \le u \le t$.

COROLLARY 4.1.– The conditional expectation of the variance can be expressed as

$$\mathbf{E}\{\sigma^2(x_t)|\,\mathcal{F}_u\} = e^{(t-u)Q}\sigma^2(x_u) \qquad [4.21]$$

for any $0 \le u \le t$.

PROPOSITION 4.7.– Suppose that $\sigma^2 \in Domain\,(Q)$, then

$$\mathbf{E}\{\sigma^4(x_t)\} = e^{tQ}\sigma^4(x) \quad \text{for } t \in \mathbb{R}_+; \tag{4.22}$$

here we have denoted $x_0 =: x$.

DEMONSTRATION.– If $\sigma^2 \in Domain\,(Q)$, then from proposition 4.1 we have that

$$m_t^{\sigma^2} := \sigma^2(x_t) - \sigma^2(x) - \int_0^t Q\,\sigma^2(x_s)ds \tag{4.23}$$

is a zero-mean martingale with respect to $\mathcal{F}_t := \sigma\{x_s; 0 \leq s \leq t\}$, where Q is the infinitesimal generator of the Markov process $(x_t)_{t\in[0,T]}$. Then σ^2 satisfies the following stochastic differential equation:

$$d\sigma^2(x_t) = Q\sigma^2(x_t)dt + dm_t^{\sigma^2}. \tag{4.24}$$

By applying Itô's lemma, we obtain

$$d\sigma^4(x_t) = 2\sigma^2(x_t)d\sigma^2(x_t) + d\langle m_\cdot^{\sigma^2}\rangle_t. \tag{4.25}$$

We note that the quadratic variation of m^{σ^2} is (see proposition 4.2) given by

$$\langle m^{\sigma^2}\rangle_t := \int_0^t [Q\sigma^4(x_s) - 2\sigma^2(x_s)Q\sigma^2(x_s)]ds. \tag{4.26}$$

Substituting the expression for the quadratic variation $\langle m^{\sigma^2}\rangle$ and for σ^2, we obtain that σ^4 satisfies the following stochastic differential equation:

$$d\sigma^4(x_t) = Q\sigma^4(x_t)dt + 2\sigma^2(x_t)dm_t^{\sigma^2}. \tag{4.27}$$

By taking the expectation at both sides of this equation, we obtain

$$\mathbf{E}\{\sigma^4(x_t)\} = \sigma^4(x) + \int_0^t Q\,\mathbf{E}\{\sigma^4(x_s)\}ds. \tag{4.28}$$

After solving this differential equation, we have the final result

$$\mathbf{E}\{\sigma^4(x_t)\} = e^{tQ}\sigma^4(x). \tag{4.29}$$

This Markov-modulated financial market is incomplete (see Elliott and Swishchuk 2007). In order to price the swaps, we need to use the minimal martingale measure. However, for simplicity we do not discuss the details regarding the minimal martingale measure, and instead focus on the pricing of future contracts on variance and volatility.

4.3.1. *Pricing variance swaps*

Let us start from the most straightforward variance swap. Variance swaps are forward contracts on future realized levels of variance. The payoff of a variance swap with expiration date T is given by

$$N(\sigma_R^2(x) - K_{var}) \qquad [4.30]$$

where $\sigma_R^2(x)$ is the realized stock variance over the life of the contract defined by

$$\sigma_R^2(x) := \frac{1}{T}\int_0^T \sigma^2(x_s)ds, \qquad [4.31]$$

where K_{var} is the strike price for variance and N is the notional amount of dollars per annualized variance point. Without loss of generality, we can assume $N = 1$. The price of the variance swap is the expected present value of the payoff in the risk-neutral world, i.e.

$$P_{var}(x) = \mathbf{E}\{e^{-rT}(\sigma_R^2(x) - K_{var})\}. \qquad [4.32]$$

The following result concerns the evaluation of the variance swap. We refer to Elliott and Swishchuk (2007) for a complete discussion and proof.

THEOREM 4.1 (Elliott and Swishchuk 2007).– The present value of a variance swap for Markov stochastic volatility is given as

$$P_{var}(x) = e^{-rT}\left\{\frac{1}{T}\int_0^T (e^{tQ}\sigma^2(x) - K_{var})dt\right\}. \qquad [4.33]$$

4.3.2. *Pricing volatility swaps*

Volatility swaps are forward contracts on future realized levels of volatility. The payoff of a volatility swap with maturity date T is given by

$$N(\sigma_R(x) - K_{vol}) \qquad [4.34]$$

where $\sigma_R(x)$ is the realized stock volatility over the life of the contract defined by

$$\sigma_R(x) := \sqrt{\frac{1}{T}\int_0^T \sigma^2(x_s)ds}, \qquad [4.35]$$

and K_{vol} is the strike price for volatility and N is the notional amount of dollars per annualized volatility point. We will assume, as before, that $N = 1$ for the sake of simplicity. The price of the volatility swap is the expected present value of the payoff in the risk-neutral world, i.e.

$$P_{vol}(x) = \mathbf{E}\{e^{-rT}(\sigma_R(x) - K_{vol})\}. \qquad [4.36]$$

For the evaluation of volatility swaps, we need to know the expected value of the square root of the variance, but unfortunately, we are not able to analytically evaluate this expected value. Then, in order to obtain a closed formula for the price of volatility swaps, we have to make an approximation. Following the approach of Brockhaus and Long (2000) (see also Javaheri *et al.* 2002), and from the second-order Taylor expansion, we have

$$\mathbf{E}\left\{\sqrt{\sigma_R^2(x)}\right\} \approx \sqrt{\mathbf{E}\{\sigma_R^2(x)\}} - \frac{Var\{\sigma_R^2(x)\}}{8\left[\mathbf{E}\{\sigma_R^2(x)\}\right]^{3/2}}. \qquad [4.37]$$

Then, in order to evaluate the volatility swap price we have to know both expectation and variance of $\sigma_R^2(x)$.

THEOREM 4.2.– The value of a volatility swap for Markov-modulated stochastic volatility is

$$P_{vol}(x) \approx e^{-rT}\left\{\sqrt{\frac{1}{T}\int_0^T e^{tQ}\sigma^2(x)dt} - \frac{Var\{\sigma_R^2(x)\}}{8\left(\frac{1}{T}\int_0^T e^{tQ}\sigma^2(x)dt\right)^{3/2}} - K_{vol}\right\},$$

where the variance is given by

$$Var\{\sigma_R^2(x)\} = \frac{2}{T^2}\int_0^T\int_0^t e^{sQ}\left[\sigma^2(x)e^{(t-s)Q}\sigma^2(x)\right]dsdt$$
$$-\left[\int_0^T e^{tQ}\sigma^2(x)dt\right]^2.$$

DEMONSTRATION.– We have already obtained the expectation of the realized variance,

$$\mathbf{E}\{\sigma_R^2(x)\} = \frac{1}{T}\int_0^T e^{tQ}\sigma^2(x)dt \qquad [4.38]$$

then it remains to prove that

$$Var\{\sigma_R^2(x)\} = \frac{2}{T^2}\int_0^T \int_0^t e^{sQ}\left[\sigma^2(x)e^{tQ}\sigma^2(x)\right] dsdt - \left[\int_0^T e^{tQ}\sigma^2(x)dt\right]^2.$$

The variance is, from the definition, given by

$$Var\{\sigma_R^2(x)\} = \mathbf{E}\{[\sigma_R^2(x) - \mathbf{E}\{\sigma_R^2(x)\}]^2\}, \qquad [4.39]$$

using the definition of realized variance, and Fubini theorem, we have

$$Var\{\sigma_R^2(x)\} = \mathbf{E}\left\{\left[\frac{1}{T}\int_0^T \sigma^2(x_t)dt - \frac{1}{T}\int_0^T \mathbf{E}\{\sigma^2(x_t)\}dt\right]^2\right\}$$

$$= \mathbf{E}\left\{\left[\frac{1}{T}\int_0^T \left(\sigma^2(x_t) - \mathbf{E}\{\sigma^2(x_t)\}\right) dt\right]^2\right\}$$

$$= \frac{1}{T^2}\int_0^T \int_0^T \mathbf{E}\left\{[\sigma^2(x_t) - \mathbf{E}\{\sigma^2(x_t)\}][\sigma^2(x_s) - \mathbf{E}\{\sigma^2(x_s)\}]\right\} dsdt, \quad [4.40]$$

and then solving the product

$$Var\{\sigma_R^2(x)\} = \frac{1}{T^2}\int_0^T \int_0^T \mathbf{E}\left\{\sigma^2(x_t)\sigma^2(x_s)\right\} dsdt$$

$$- \left[\frac{1}{T}\int_0^T \mathbf{E}\{\sigma^2(x_t)\}dt\right]^2. \qquad [4.41]$$

The second term on the right-hand side is known; it follows directly from proposition 4.6. Moreover, we observe that the integrand on the first term is invariant in the exchange of s and t. Then, we can divide the integration set into the two zones above and below the line $t = s$. According to the symmetry of the problem, the contribution on the two parts is the same. Thus, we can rewrite the variance as

$$Var\{\sigma_R^2(x)\} = \frac{2}{T^2}\int_0^T \int_0^t \mathbf{E}\left\{\sigma^2(x_t)\sigma^2(x_s)\right\} dsdt$$

$$-\left[\frac{1}{T}\int_0^T e^{tQ}\sigma^2(x)dt\right]^2. \qquad [4.42]$$

We stress that, in this form, the integration set of the first term is such that the inequality $s \leq t$ holds true. Using the properties of conditional expectation, we have

$$Var\{\sigma_R^2(x)\} = \frac{2}{T^2}\int_0^T\int_0^t \mathbf{E}\{\sigma^2(x_s)\mathbf{E}\{\sigma^2(x_t)|\,\mathcal{F}_s\}\}dsdt$$

$$-\left[\frac{1}{T}\int_0^T e^{tQ}\sigma^2(x)dt\right]^2. \qquad [4.43]$$

Using the Markov property and corollary 4.1, the conditional expected value in the integrand can be viewed as a function of the process at time s, i.e.

$$\mathbf{E}\{\sigma^2(x_t)|\,\mathcal{F}_s\} = e^{(t-s)Q}\sigma(x_s) =: g(x_s). \qquad [4.44]$$

Thus, the variance can be expressed as

$$Var\{\sigma_R^2(x)\} = \frac{2}{T^2}\int_0^T\int_0^t \mathbf{E}\{\sigma^2(x_s)g(x_s)\}dsdt - \left[\frac{1}{T}\int_0^T e^{tQ}\sigma^2(x)dt\right]^2.$$

Now, using proposition 4.5, we can solve the expectation on the integrand and we finally obtain

$$Var\{\sigma_R^2(x)\} = \frac{2}{T^2}\int_0^T\int_0^t e^{sQ}\left(\sigma^2(x)g(x)\right)dsdt - \left[\frac{1}{T}\int_0^T e^{tQ}\sigma^2(x)dt\right]^2.$$

Moreover, we observe that function g, evaluated in x, becomes

$$g(x) = e^{(t-s)Q}\sigma^2(x). \qquad [4.45]$$

Then, substituting in the previous formula we can express the variance as

$$Var\{\sigma_R^2(x)\} = \frac{2}{T^2}\int_0^T\int_0^t e^{sQ}\left[\sigma^2(x)e^{(t-s)Q}\sigma^2(x)\right]dsdt$$

$$-\left[\frac{1}{T}\int_0^T e^{tQ}\sigma^2(x)dt\right]^2. \qquad [4.46]$$

4.4. Covariance and correlation swaps for two risky assets for financial markets with Markov-modulated stochastic volatilities

Let us consider a market model with two risky assets and a risk-free bond. Let us assume that the risky assets satisfy the following stochastic differential equations:

$$\begin{cases} dS_t^{(1)} = S_t^{(1)}(\mu_t^{(1)}dt + \sigma^{(1)}(x_t)dw_t^{(1)}) \\ dS_t^{(2)} = S_t^{(2)}(\mu_t^{(2)}dt + \sigma^{(2)}(x_t)dw_t^{(2)}) \end{cases} \quad [4.47]$$

where $\mu^{(1)}$, $\mu^{(2)}$ are deterministic functions of time and $(w_t^{(1)})_t$ and $(w_t^{(2)})_t$ are standard Wiener processes with quadratic covariance given by

$$d[w_t^{(1)}, w_t^{(2)}] = \rho_t dt. \quad [4.48]$$

Here, ρ_t is a deterministic function of time and $(w_t^{(1)})_t$, $(w_t^{(2)})_t$ are supposed to be independent of the Markov process $(x_t)_t$. This model allows us to study the covariance and correlation structure of two risky assets and how it is possible to price future contracts on them.

4.4.1. *Pricing covariance swaps*

A covariance swap is a covariance forward contract on the realized covariance between two risky assets, whose payoff at maturity is equal to

$$N(Cov_R(S^{(1)}, S^{(2)}) - K_{cov}) \quad [4.49]$$

where K_{cov} is a strike reference value, N is the notional amount and $Cov_R(S^{(1)}, S^{(2)})$ is the realized covariance of the two assets $S^{(1)}$ and $S^{(2)}$ defined by

$$Cov_R(S^{(1)}, S^{(2)}) = \frac{1}{T}\left[\ln S_T^{(1)}, \ln S_T^{(2)}\right]$$

$$= \frac{1}{T}\int_0^T \rho_t \sigma^{(1)}(x_t)\sigma^{(2)}(x_t)dt. \quad [4.50]$$

The price of the covariance swap is the expected present value of the payoff in the risk neutral world

$$P_{cov}(x) = \mathbf{E}\{e^{-rT}(Cov_R(S^{(1)}, S^{(2)}) - K_{cov})\}, \quad [4.51]$$

where we assumed that $N = 1$.

THEOREM 4.3.– The value of a covariance swap for Markov-modulated stochastic volatility is

$$P_{cov}(x) = e^{-rT}\left\{\frac{1}{T}\int_0^T \rho_t e^{tQ}[\sigma^{(1)}(x)\sigma^{(2)}(x)]dt - K_{cov}\right\}. \quad [4.52]$$

DEMONSTRATION.– To evaluate the price of a covariance swap, we only need to know

$$\mathbf{E}\{Cov_R(S^{(1)},S^{(2)})\} = \frac{1}{T}\int_0^T \rho_t\, \mathbf{E}\{\sigma^{(1)}(x_t)\sigma^{(2)}(x_t)\}dt. \quad [4.53]$$

Then, it remains to show that

$$\mathbf{E}\{\sigma^{(1)}(x_t)\sigma^{(2)}(x_t)\} = e^{tQ}[\sigma^{(1)}(x)\sigma^{(2)}(x)]. \quad [4.54]$$

By applying Itô's lemma, we have

$$\begin{aligned} d(\sigma^{(1)}(x_t)\sigma^{(2)}(x_t)) &= \sigma^{(1)}(x_t)d\sigma^{(2)}(x_t) + \sigma^{(2)}(x_t)d\sigma^{(1)}(x_t) \\ &\quad + d\langle\sigma^{(1)}(x.),\sigma^{(2)}(x.)\rangle_t. \end{aligned} \quad [4.55]$$

Using proposition 4.3, we can express the covariation as

$$\begin{aligned} d\langle\sigma^{(1)}(x.),\sigma^{(2)}(x.)\rangle_t &= Q[\sigma^{(1)}(x_t)\sigma^{(2)}(x_t)]dt - [\sigma^{(1)}(x_t)Q\sigma^{(2)}(x_t) \\ &\quad + \sigma^{(2)}(x_t)Q\sigma^{(1)}(x_t)]dt. \end{aligned} \quad [4.56]$$

Furthermore, from proposition 4.1, we have

$$d\sigma^{(i)}(x_t) = Q\sigma^{(i)}(x_t)dt + dm^{\sigma^{(i)}} \qquad i = 1, 2. \quad [4.57]$$

Substituting [4.56] and [4.57] into equation [4.55] gives

$$\begin{aligned} d(\sigma^{(1)}(x_t)\sigma^{(2)}(x_t)) &= Q[\sigma^{(1)}(x_t)\sigma^{(2)}(x_t)]dt + \sigma^{(1)}(x_t)dm^{\sigma^{(2)}} \\ &\quad + \sigma^{(2)}(x_t)dm^{\sigma^{(1)}}. \end{aligned} \quad [4.58]$$

Now, by taking the expectation on both sides of the above equation we obtain

$$\begin{aligned} \mathbf{E}\{\sigma^{(1)}(x_t)\sigma^{(2)}(x_t)\} &= \sigma^{(1)}(x)\sigma^{(2)}(x) \\ &\quad + \int_0^t Q\mathbf{E}\{\sigma^{(1)}(x_s)\sigma^{(2)}(x_s)\}dt. \end{aligned} \quad [4.59]$$

Now, we solve this differential equation, giving the solution

$$\mathbf{E}\{\sigma^{(1)}(x_t)\sigma^{(2)}(x_t)\} = e^{tQ}[\sigma^{(1)}(x)\sigma^{(2)}(x)], \quad [4.60]$$

and this concludes the proof.

4.4.2. *Pricing correlation swaps*

A correlation swap is a forward contract on the correlation between the underlying assets S^1 and S^2, whose payoff at maturity is equal to

$$N(Corr_R(S^1, S^2) - K_{corr}) \tag{4.61}$$

where K_{corr} is a strike reference level, N is the notional amount and the realized correlation $Corr_R(S^1, S^2)$ is defined by

$$Corr_R(S^1, S^2) = \frac{Cov_R(S^1, S^2)}{\sqrt{\sigma_R^{(1)^2}(x)}\sqrt{\sigma_R^{(2)^2}(x)}}, \tag{4.62}$$

where the realized variance is given by

$$\sigma_R^{(i)^2}(x) = \frac{1}{T}\int_0^T (\sigma^{(i)}(x_t))^2 dt \qquad i = 1, 2. \tag{4.63}$$

The price of the correlation swap is the expected present value of the payoff in the risk neutral world

$$P_{corr}(x) = \mathbf{E}\{e^{-rT}(Corr_R(S^1, S^2) - K_{corr})\} \tag{4.64}$$

where we set $N = 1$ for simplicity. Unfortunately, the expected value of $Corr_R(S^1, S^2)$ is not known analytically. In order to obtain an explicit formula for the correlation swap price, we need to introduce an approximation.

4.4.3. *Correlation swap made simple*

First of all, let us introduce the following notations:

$$\begin{aligned} X &= Cov_R(S^1, S^2) \\ Y &= \sigma_R^{(1)^2}(x) \\ Z &= \sigma_R^{(2)^2}(x); \end{aligned} \tag{4.65}$$

in the following formulas with the subscript 0, we will denote the expected value of the above random variables. Starting from the approach we have used for the volatility swap, we would like to approximate the square root of Y and Z at the first order as follows:

$$\begin{aligned}\sqrt{Y} &\approx \sqrt{Y_0} + \frac{Y - Y_0}{2\sqrt{Y_0}}\\ \sqrt{Z} &\approx \sqrt{Z_0} + \frac{Z - Z_0}{2\sqrt{Z_0}}.\end{aligned} \qquad [4.66]$$

The realized correlation can now be approximated by

$$\begin{aligned}Corr_R(S^1, S^2) &\approx \frac{X}{\left(\sqrt{Y_0} + \frac{Y-Y_0}{2\sqrt{Y_0}}\right)\left(\sqrt{Z_0} + \frac{Z-Z_0}{2\sqrt{Z_0}}\right)}\\ &= \frac{\frac{X}{\sqrt{Y_0}\sqrt{Z_0}}}{\left(1 + \frac{Y-Y_0}{2Y_0}\right)\left(1 + \frac{Z-Z_0}{2Z_0}\right)}.\end{aligned} \qquad [4.67]$$

Solving the product in the denominator on the right-hand side last term and only keeping the terms up to the first order in the increment, we have

$$\begin{aligned}Corr_R(S^1, S^2) &\approx \frac{\frac{X}{\sqrt{Y_0}\sqrt{Z_0}}}{1 + \left(\frac{Y-Y_0}{2Y_0} + \frac{Z-Z_0}{2Z_0}\right)}\\ &\approx \frac{X}{\sqrt{Y_0}\sqrt{Z_0}}\left[1 - \left(\frac{Y - Y_0}{2Y_0} + \frac{Z - Z_0}{2Z_0}\right)\right].\end{aligned} \qquad [4.68]$$

Finally, we obtain the following approximation for the correlation:

$$Corr_R(S^1, S^2) \approx \frac{X}{\sqrt{Y_0}\sqrt{Z_0}} - \frac{X}{\sqrt{Y_0}\sqrt{Z_0}}\left(\frac{Y - Y_0}{2Y_0} + \frac{Z - Z_0}{2Z_0}\right). \qquad [4.69]$$

We will only evaluate the expectation on the first term on the right-hand side, which represents the zero-order approximation and is the most intuitive part. We discuss the first-order correction in the Appendix. In what follows, we are going to approximate the realized correlation as

$$Corr_R(S^1, S^2) \approx \frac{X}{\sqrt{Y_0}\sqrt{Z_0}}, \qquad [4.70]$$

hence, substituting X, Y and Z we obtain

$$Corr_R(S^1, S^2) \approx \frac{1}{\sqrt{\mathbf{E}\{\sigma_R^{(1)^2}(x)\}}\sqrt{\mathbf{E}\{\sigma_{2R}^{(2)^2}(x)\}}}$$

$$\frac{1}{T}\int_0^T \rho_t \sigma^{(1)}(x_t)\sigma^{(2)}(x_t)dt \quad [4.71]$$

where (see corollary 4.1), we have

$$\mathbf{E}\left\{\sigma^2_{(i)R}(x)\right\} = \mathbf{E}\left\{\frac{1}{T}\int_0^T (\sigma^{(i)}(x_t))^2 dt\right\} = \frac{1}{T}\int_0^T e^{tQ}(\sigma^{(i)}(x))^2 dt, \quad [4.72]$$

for $i = 1, 2$. In order to price a correlation swap, we have to be able to evaluate the expectation of both sides of equation [4.71]. From proposition 4.5, the expectation of the integrand in the right-hand side of [4.71] is given by

$$\mathbf{E}\{\sigma^{(1)}(x_t)\sigma^{(2)}(x_t)\} = e^{tQ}\sigma^{(1)}(x)\sigma^{(2)}(x). \quad [4.73]$$

We can summarize the previous result in the following statement.

THEOREM 4.4.– The value of a correlation swap for a Markov-modulated stochastic volatility is given as:

$$P_{corr}(x) \approx e^{-rT}$$

$$\times \left[\frac{\int_0^T \rho_t e^{tQ}\sigma^{(1)}(x)\sigma^{(2)}(x)dt}{\sqrt{\int_0^T e^{tQ}(\sigma^{(1)}(x))^2 dt}\sqrt{\int_0^T e^{tQ}(\sigma^{(2)}(x))^2 dt}} - K_{corr}\right]. \quad [4.74]$$

4.5. Example: variance, volatility, covariance and correlation swaps for stochastic volatility driven by two state continuous Markov chain

Let $(x_t)_t$ be a two-state continuous time Markov chain, and let us denote the states as "Up" (u) and "Down" (d). Let Q be the generator of this Markov chain

$$Q = \begin{pmatrix} q_{uu} & q_{ud} \\ q_{du} & q_{dd} \end{pmatrix} \quad [4.75]$$

and let

$$P(t) = \begin{pmatrix} p_{uu}(t) & p_{ud}(t) \\ p_{du}(t) & p_{dd}(t) \end{pmatrix} \quad [4.76]$$

be its transition function, such that

$$P(t) = e^{tQ}. \quad [4.77]$$

In this simple model, the volatility only takes two values: σ_u and σ_d. Thus, we can easily express the swap prices of the future contracts studied hitherto.
The variance swap price in this model is given by

$$P_{var}(i) = e^{-rT}\left\{\frac{1}{T}\int_0^T (p_{iu}(t)\sigma_u^2 + p_{id}(t)\sigma_d^2)dt - K_{var}\right\}, \quad [4.78]$$

where $i = u, d$ is the initial state of the Markov chain. If we are uncertain about the initial state and we only have a probability distribution, let us say (p_u, p_d) such that $p_u + p_d$, then the price is going to be

$$P_{var} = p_u P_{var}(u) + p_d P_{var}(d). \quad [4.79]$$

If we assume that the initial distribution is actually the stationary distribution of this Markov chain, then the price simply becomes

$$P_{var} = e^{-rT}\left\{\pi_u\sigma_u^2 + \pi_d\sigma_d^2 - K_{var}\right\}, \quad [4.80]$$

where (π_u, π_d) is the stationary distribution.

The volatility swap price in this model can be approximated by

$$\begin{aligned} P_{vol}(i) \approx\ & e^{-rT}\Bigg\{\sqrt{\frac{1}{T}\int_0^T [p_{iu}(t)\sigma_u^2 + p_{id}(t)\sigma_d^2]dt} \\ & - \frac{Var\{\sigma_R^2(i)\}}{8\left(\frac{1}{T}\int_0^T [p_{iu}(t)\sigma_u^2 + p_{id}(t)\sigma_d^2]dt\right)^{3/2}} - K_{vol}\Bigg\} \end{aligned} \quad [4.81]$$

for $i = u, d$ and where

$$\begin{aligned} Var\{\sigma_R^2(i)\} = \frac{2}{T^2}\int_0^T\int_0^t & \Big\{p_{iu}(s)p_{uu}(t-s)\sigma_u^4 \\ & + [p_{iu}(s)p_{ud}(t-s) + p_{id}(s)p_{du}(t-s)]\sigma_d^2\sigma_u^2 \\ & + p_{id}(s)p_{dd}(t-s)\sigma_d^4\Big\}dsdt \\ & - \left[\frac{1}{T}\int_0^T [p_{iu}(t)\sigma_u^2 + p_{id}(t)\sigma_d^2]dt\right]^2. \end{aligned}$$

Let us now consider a two-risky-asset market model with volatility modulated by this two-state Markov chain. In this setting, the covariance swap price is

$$P_{cov}(i) = e^{-rT} \times \left\{ \frac{1}{T} \int_0^T \rho_t [p_{iu}(t)\sigma_u^{(1)}\sigma_u^{(2)} + p_{id}(t)\sigma_d^{(1)}\sigma_d^{(2)}]dt - K_{cov} \right\}, \qquad [4.82]$$

for $i = u, d$ representing the initial state of the chain.

The correlation swap in this model can be approximated by

$$P_{corr}(i) \approx e^{-rT} \left[\frac{M_1}{\sqrt{M_2}\sqrt{M_3}} - K_{corr} \right]$$

for $i = u, d$ standing for the initial state of the chain. The constants M_1, M_2 and M_3 are defined as

$$M_1 = \int_0^T \rho_t [p_{iu}(t)\sigma_u^{(1)}\sigma_u^{(2)} + p_{id}(t)\sigma_d^{(1)}\sigma_d^{(2)}]dt,$$

$$M_2 = \int_0^T [p_{iu}(t)(\sigma_u^{(1)})^2 + p_{id}(t)(\sigma_d^{(1)})^2]dt,$$

$$M_3 = \int_0^T [p_{iu}(t)(\sigma_u^{(2)})^2 + p_{id}(t)(\sigma_d^{(2)})^2]dt.$$

4.6. Numerical example

4.6.1. *S&P 500: variance and volatility swaps*

In this section, we use the two-state continuous Markov chain model for stochastic volatility, to price variance and volatility swaps on the volatility of the S&P 500 index. We used the CBOE Volatility Index (VIX) (January 2004 to June 2012).

The daily data was used to construct the one-step transition matrix for the two-state Markov chain. In particular, we used the interpolation between the highest and the lowest daily data. We calculated the mean value of the interpolated data and we divided the data in up, higher of the mean, and down, lower or equal to the mean value. In Table 4.1, we show the one-step transition probability matrix.

Using this probability matrix, given the initial state of the chain, we are able to evaluate the price of the covariance and correlation swap, as described in section 4.5. The variance and volatility swap prices, as a function of maturity, are shown in Figure 4.1. The rate of convergence of the prices depends on the rate of convergence of the transition probability matrix to the stationary distribution.

Transition matrix		
	Up	Down
Up	0.957	0.043
Down	0.026	0.974

Table 4.1. *One-step transition probability matrix*

4.6.2. *S&P 500 and NASDAQ-100: covariance and correlation swaps*

In this section, we use the two-state continuous Markov chain model for stochastic volatility, to price covariance and correlation swaps. The underlying assets are the S&P 500 index, the CBOE Volatility Index (VIX), the NASDAQ-100 index and the CBOE NASDAQ-100 Volatility Index (VXN) (January 2004 to June 2012).

The daily data was used to construct the one-step transition matrix for the two-state Markov chain. In particular, we used the interpolation between the highest and the lowest daily data. We calculated the mean value of the interpolated data and we divided the data in up, higher of the mean, and down, lower or equal to the mean value. In Table 4.2, we show the one-step transition probability matrix built using the VIX and VXN indexes.

Transition matrix		
	Up	Down
Up	0.950	0.050
Down	0.030	0.970

Table 4.2. *One-step transition probability matrix*

Using this probability matrix, given the initial state of the chain, we are able to evaluate the price of covariance and correlation swaps as described in section 4.5. In the numerical evaluation, we assumed that the correlation between the Brownian motion ρ (see equation [4.48]) was constant.

The covariance and correlation swap prices, as a function of maturity, are shown in Figure 4.2. The rate of convergence of the prices depends on the rate of convergence of the transition probability matrix to the stationary distribution. For the correlation swap price, we show the result up to the zero order of approximation, as described in section 4.5, and up to the first order of approximation, as described in the Appendix. The zero order of the approximation price is constant. Indeed, at that order of approximation, we only take into account the correlation of the asset noise, that is ρ, the Brownian motions correlation. In the first order, the interaction of the volatilities, due to the Markov process, is taken into account and the price changes as a function of maturity.

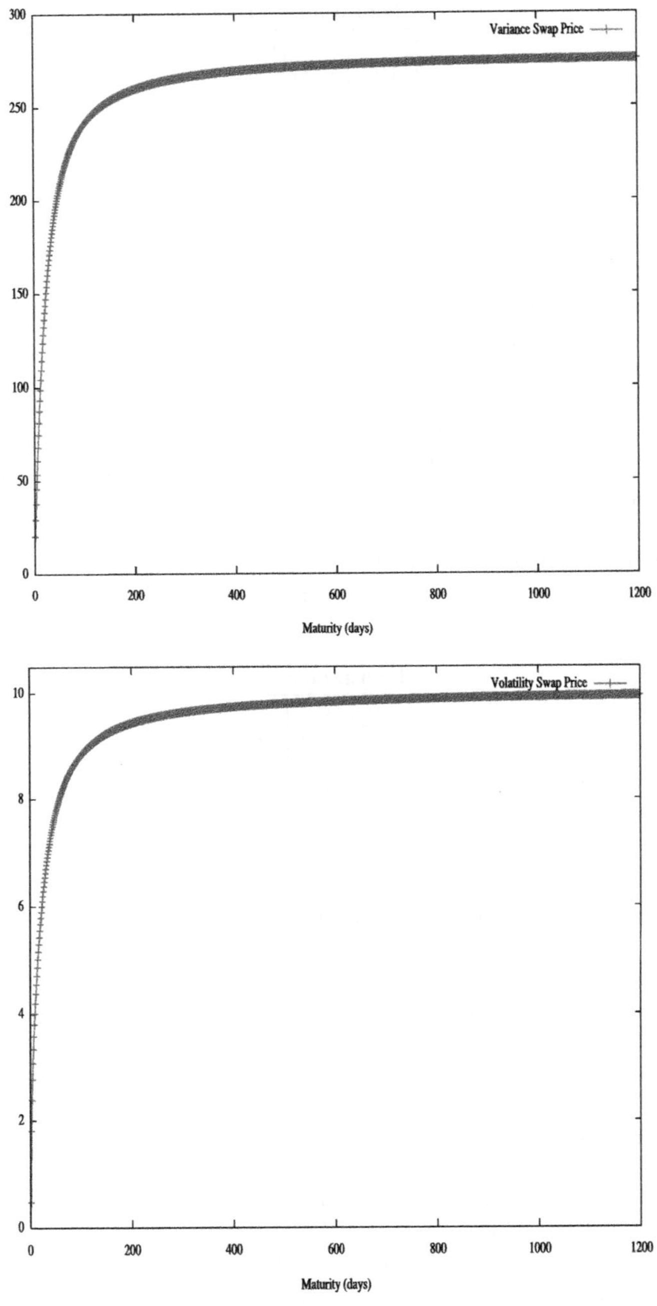

Figure 4.1. *Variance and volatility swap prices . For a color version of this figure, see www.iste.co.uk/pogorui/random2.zip*

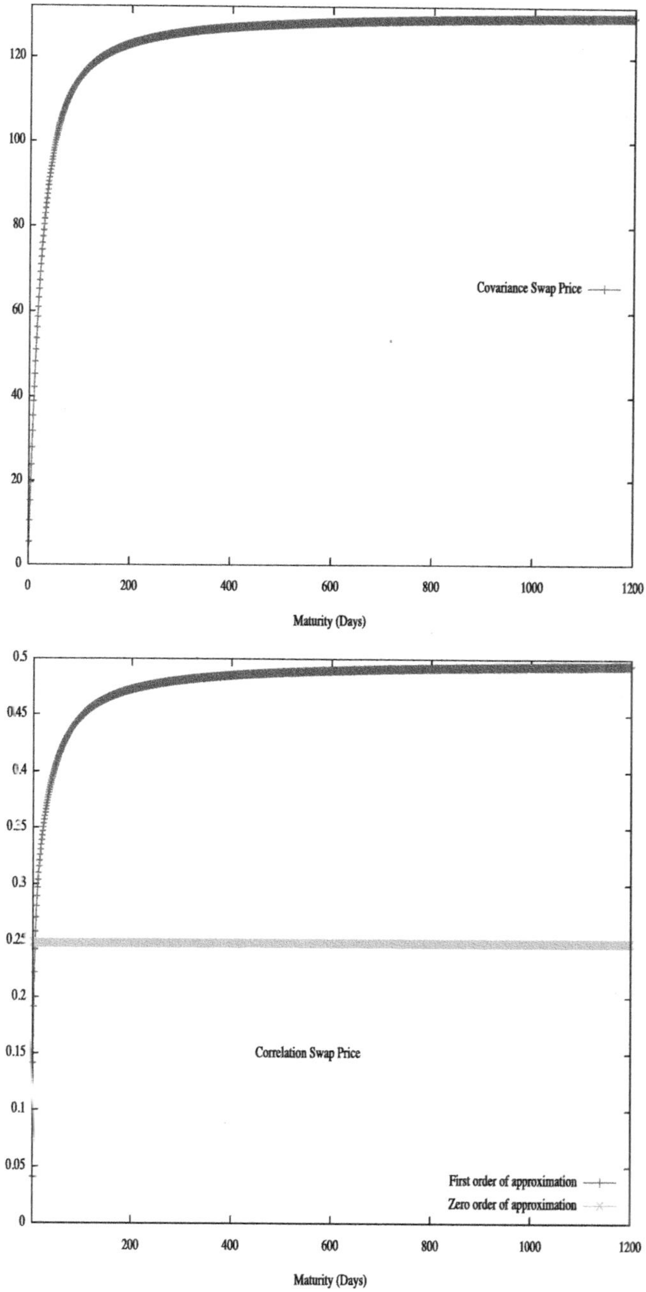

Figure 4.2. *Variance and volatility swap prices . For a color version of this figure, see www.iste.co.uk/pogorui/random2.zip*

4.7. Appendix 1

4.7.1. *Correlation swaps: first-order correction*

We would like to obtain an approximation for the realized correlation between two risky assets

$$Corr_R(S^1, S^2) = \frac{Cov_R(S^1, S^2)}{\sqrt{\sigma_R^{(1)^2}(x)}\sqrt{\sigma_R^{(2)^2}(x)}}. \qquad [4.83]$$

In section 4.4.3, we have already obtained the following approximated expression:

$$Corr_R(S^1, S^2) \approx \frac{\frac{X}{\sqrt{Y_0}\sqrt{Z_0}}}{1 + \left(\frac{Y-Y_0}{2Y_0} + \frac{Z-Z_0}{2Z_0}\right)}$$
$$\approx \frac{X}{\sqrt{Y_0}\sqrt{Z_0}}\left[1 - \left(\frac{Y-Y_0}{2Y_0} + \frac{Z-Z_0}{2Z_0}\right)\right], \qquad [4.84]$$

where

$$\begin{aligned} X &= Cov_R(S^1, S^2) \\ Y &= \sigma_R^{(1)^2}(x) \\ Z &= \sigma_R^{(2)^2}(x), \end{aligned} \qquad [4.85]$$

and with the subscript 0, we have denoted the expected values. We have already evaluated the expectation of the zero-order approximation: now, we would like to evaluate the first order.

Substituting X, Y and Z into equation [4.84], we obtain

$$Corr_R(S^1, S^2) \approx \frac{1}{\sqrt{\mathbf{E}\{\sigma_R^{(1)^2}(x)\}}\sqrt{\mathbf{E}\{\sigma_{2R}^{(2)^2}(x)\}}}\frac{1}{T}\int_0^T \rho_t\sigma^{(1)}(x_t)\sigma^{(2)}(x_t)dt$$

$$-\frac{1}{2T^2(\mathbf{E}\{\sigma_R^{(1)^2}(x)\})^{3/2}(\mathbf{E}\{\sigma_R^{(2)^2}(x)\})^{3/2}}\int_0^T \rho_t\sigma^{(1)}(x_t)\sigma^{(2)}(x_t)dt$$

$$\times\Bigg\{\mathbf{E}\{\sigma_R^{(2)^2}(x)\}\int_0^T [(\sigma^{(1)}(x_s))^2 - \mathbf{E}\{(\sigma^{(1)}(x_s))^2\}]ds$$

$$+\mathbf{E}\{\sigma_R^{(1)^2}(x)\}\int_0^T[(\sigma^{(2)}(x_u))^2-\mathbf{E}\{(\sigma^{(2)}(x_u))^2\}]du\Bigg\}, \qquad [4.86]$$

where (see theorem 4.1) we have

$$\mathbf{E}\left\{\sigma_{(i)R}^2(x)\right\}=\mathbf{E}\left\{\frac{1}{T}\int_0^T(\sigma^{(i)}(x_t))^2dt\right\}=\frac{1}{T}\int_0^T e^{tQ}(\sigma^{(i)}(x))^2dt, \qquad [4.87]$$

for $i=1,2$. We have to evaluate the expectation of the right-hand side of equation [4.86]. In section 4.4.3, we calculated the expectation of the first term. Now, we will focus on the other terms. First of all, let us rewrite them as follows:

$$\int_0^T\int_0^T\rho_t\sigma^{(1)}(x_t)\sigma^{(2)}(x_t)\Big(\mathbf{E}\{\sigma_R^{(2)^2}(x)\}[(\sigma^{(1)}(x_s))^2-\mathbf{E}\{(\sigma^{(1)}(x_s))^2\}]$$

$$+\mathbf{E}\{\sigma_R^{(\cdot)^2}(x)\}[(\sigma^{(2)}(x_s))^2-\mathbf{E}\{(\sigma^{(2)}(x_s))^2\}]\Big)dsdt. \qquad [4.88]$$

We have four different contributions in the integrals, and the expectation of the terms

$$\int_0^T\int_0^T\rho_t\sigma^{(1)}(x_t,\gamma(t))\sigma^{(2)}(x_t)\mathbf{E}\{\sigma^{(i)^2}(x_s)\}\mathbf{E}\{\sigma^{(-i)^2}(x_s)\}dsdt \qquad [4.89]$$

for $i=1,2$, can be evaluate using theorem 4.3. Then, in order to evaluate the expectation of the approximated realized correlation, it only remains to calculate

$$\mathbf{E}\left\{\int_0^T\int_0^T\rho_t\,\sigma^{(1)}(x_t)\,\sigma^{(2)}(x_t)\,\sigma^{(i)^2}(x_s)dsdt\right\} \qquad i=1,2. \qquad [4.90]$$

To this end, let us first divide the range of integration into two intervals as follows:

$$\mathbf{E}\Bigg\{\int_0^T\int_0^t\rho_t\,\sigma^{(1)}(x_t)\,\sigma^{(2)}(x_t)\,\sigma^{(i)^2}(x_s)dsdt$$

$$+\int_0^T\int_t^T\rho_t\,\sigma^{(1)}(x_t)\,\sigma^{(2)}(x_t)\,\sigma^{(i)^2}(x_s)dsdt\Bigg\} \qquad [4.91]$$

for $i=1,2$. We note that the first integral set is such that $t>s$, whereas the second has $t<s$. Now, we can use the property of conditional expectation to obtain

$$\mathbf{E}\Bigg\{\int_0^T\int_0^t\rho_t\,\mathbf{E}\{\sigma^{(1)}(x_t)\,\sigma^{(2)}(x_t)|\,\mathcal{F}_s\}\,\sigma^{(i)^2}(x_s)dsdt$$

$$+\int_0^T\int_t^T\rho_t\,\sigma^{(1)}(x_t)\,\sigma^{(2)}(x_t)\mathbf{E}\{\sigma^{(i)^2}(x_s)|\,\mathcal{F}_t\}dsdt\Bigg\}. \qquad [4.92]$$

Using the Markov property, we can express the conditional expectations as

$$\mathbf{E}\{\sigma^{(1)}(x_t)\sigma^{(2)}(x_t)|\,\mathcal{F}_s\} = e^{(t-s)Q}\sigma^{(1)}(x_s)\sigma^{(2)}(x_s) =: h(x_s) \quad [4.93]$$

for $t > s$ and

$$\mathbf{E}\{\sigma^{(i)^2}(x_s)|\,\mathcal{F}_t\} = e^{(s-t)Q}\sigma^{(i)^2}(x_t) =: g^{(i)}(x_t) \quad [4.94]$$

for $s > t$. Therefore, the first term of equation [4.92] can be expressed (see proposition 4.5) as

$$\mathbf{E}\left\{\int_0^T \int_0^t \rho_t h(x_s)\sigma^{(i)^2}(x_s)dsdt\right\}$$

$$= \int_0^T \int_0^t \rho_t e^{sQ}[h(x)\sigma^{(i)^2}(x)]dsdt, \quad [4.95]$$

whereas the second term can be expressed as

$$\mathbf{E}\left\{\int_0^T \int_t^T \rho_t \sigma^{(1)}(x_t)\sigma^{(2)}(x_t)g^{(i)}(x_t)dsdt\right\}$$
$$= \int_0^T \int_t^T \rho_t e^{tQ}[\sigma^{(1)}(x)\sigma^{(2)}(x)g^{(i)}(x)]dsdt. \quad [4.96]$$

Now, we evaluate the functions h and g at x obtaining

$$h(x) = e^{(t-s)Q}[\sigma^{(1)}(x)\sigma^{(2)}(x)] \quad [4.97]$$

and

$$g^{(i)}(x) = e^{(s-t)Q}[\sigma^{(i)^2}(x)]. \quad [4.98]$$

We can summarize the previous result in the following statement, which gives the correlation swap price up to the first order of approximation.

THEOREM 4.5.– The value of the correlation swap for a Markov-modulated volatility is

$$P_{corr}(x) = e^{-rT}\left(\mathbf{E}\{Corr_R(S^1, S^2)\} - K_{corr}\right) \quad [4.99]$$

where the realized correlation can be approximated by

$$
\begin{aligned}
\mathbf{E}\{Corr_R(S^1,S^2)\} &\approx \frac{2\int_0^T \rho_t e^{tQ}\sigma^{(1)}(x)\sigma^{(2)}(x)dt}{\sqrt{K_1}\sqrt{K_2}} \\
&-\frac{1}{2K_1^{3/2}K_2^{1/2}}\Bigg\{\int_0^T \rho_t\Bigg(\int_0^t e^{sQ}\left\{e^{(t-s)Q}[\sigma^{(1)}(x)\sigma^{(2)}(x)]\sigma^{(1)^2}(x)\right\}ds \\
&+\int_t^T e^{tQ}\left\{\sigma^{(1)}(x)\sigma^{(2)}(x)e^{(u-t)Q}[\sigma^{(1)^2}(x)]\right\}du\Bigg)dt\Bigg\} \qquad [4.100] \\
&-\frac{1}{2K_1^{1/2}K_2^{3/2}}\Bigg\{\int_0^T \rho_t\Bigg(\int_0^t e^{sQ}\left\{e^{(t-s)Q}[\sigma^{(1)}(x)\sigma^{(2)}(x)]\sigma^{(2)^2}(x)\right\}ds \\
&+\int_t^T e^{tQ}\left\{\sigma^{(1)}(x)\sigma^{(2)}(x)e^{(u-t)Q}[\sigma^{(2)^2}(x)]\right\}du\Bigg)dt\Bigg\},
\end{aligned}
$$

where

$$
K_1 = \int_0^T e^{tQ}(\sigma^{(1)}(x))^2 dt
$$

$$
K_2 = \int_0^T e^{tQ}(\sigma^{(2)}(x))^2 dt.
$$

5

Modeling and Pricing of Variance, Volatility, Covariance and Correlation Swaps for Financial Markets with Semi-Markov Volatilities

In this chapter, we model financial markets with semi-Markov volatilities and price variance, volatility, covariance and correlation swaps. Numerical evaluations of variance, volatility, covariance and correlations swaps with semi-Markov volatility are also presented. The novelty of the chapter lies in the pricing of volatility swaps in the closed form, and the pricing of covariance and correlation swaps in a market with two risky assets. The results of this chapter are based mainly on the working paper by Salvi and Swishchuk (2012b).

5.1. Introduction

Some of the newest and most recent financial products are variance and volatility swaps, which are useful for volatility hedging and speculation. The market for variance and volatility swaps has been growing, and many investment banks and other financial institutions are now actively quoting volatility swaps on various assets: stock indexes, currencies, as well as commodities. A stock's volatility is the simplest measure of its riskiness or uncertainty. Formally, the volatility σ_R is the annualized standard deviation of the stock's returns during the period of interest, where the subscript R denotes the observed or "realized" volatility. Why trade volatility or variance? As mentioned in Arfken *et al.* (2012):

> [...] just as stock investors think they know something about the direction of the stock market, so we may think we have insight into the level of

future volatility. If we think current volatility is low, for the right price we might want to take a position that profits if volatility increases.

In this chapter, we model financial markets with semi-Markov volatilities, price covariance and correlation swaps. Numerical evaluations of variance, volatility, covariance and correlations swaps with semi-Markov volatility are also presented.

Volatility swaps are forward contracts on future realized stock volatility, while variance swaps are similar contracts on variance, the square of the future volatility. Both of these instruments provide an easy manner for investors to gain exposure to the future level of volatility. A stock's volatility is the simplest measure of its riskless or uncertainty. Formally, the volatility σ_R is the annualized standard deviation of the stock's returns during the period of interest, where the subscript R denotes the observed or "realized" volatility.

A straightforward procedure to trade volatility is to use volatility swaps, sometimes called realized volatility forward contracts, because they provide pure exposure to volatility (and only to volatility).

A stock *volatility swap* is a forward contract on the annualized volatility. Its payoff at expiration is equal to

$$N(\sigma_R(S) - K_{vol}),$$

where $\sigma_R(S)$ is the realized stock volatility (quoted in annual terms) over the life of the contract,

$$\sigma_R(S) := \sqrt{\frac{1}{T}\int_0^T \sigma_s^2 ds},$$

where σ_t is a stochastic stock volatility, K_{vol} is the annualized volatility delivery price and N is the notional amount of the swap in dollars per annualized volatility point. The holder of a volatility swap at expiration, receives N dollars for every point by which the stock's realized volatility σ_R has exceeded the volatility delivery price K_{vol}. The holder is swapping a fixed volatility K_{vol} for the actual (floating) future volatility σ_R. We note that usually $N = \alpha I$, where α is a converting parameter with the typical value of 1 per volatility square, and I is a long-short index (+1 for long and -1 for short).

Although options market participants talk of volatility, it is variance, or volatility squared, that has a more fundamental significance (see Demeterfi *et al.* 1999).

A variance swap is a forward contract on annualized variance, i.e. the square of the realized volatility. Its payoff at expiration is equal to

$$N(\sigma_R^2(S) - K_{var}),$$

where $\sigma_R^2(S)$ is the realized stock variance (quoted in annual terms), over the life of the contract,

$$\sigma_R^2(S) := \frac{1}{T}\int_0^T \sigma_s^2 ds,$$

K_{var} is the delivery price for variance and N is the notional amount of the swap in dollars per annualized volatility point squared. The holder of a variance swap at expiration, receives N dollars for every point by which the stock's realized variance $\sigma_R^2(S)$ has exceeded the variance delivery price K_{var}.

Therefore, pricing the variance swap reduces to calculating the realized volatility square.

Valuing a variance forward contract or swap is not different from valuing any other derivative security. The value of a forward contract $\mathbf{P}$ on future realized variance, with strike price K_{var}, is the expected present value of the future payoff in the risk-neutral world:

$$\mathbf{P} = \mathbf{E}\left\{e^{-rT}(\sigma_R^2(S) - K_{var})\right\},$$

where r is the risk-free discount rate corresponding to the expiration date T, and $\mathbf{E}$ denotes the expectation.

Thus, for calculating variance swaps we only need to know $\mathbf{E}\{\sigma_R^2(S)\}$, namely, the mean value of the underlying variance.

To calculate volatility swaps, we need something else. From Brockhaus and Long's (2000) approximation (where the second-order Taylor expansion for function $\sqrt{x}$ is used), we have (see also Javaheri *et al.* 2002):

$$\mathbf{E}\{\sqrt{\sigma_R^2(S)}\} \approx \sqrt{\mathbf{E}\{V\}} - \frac{Var\{V\}}{8\,[\mathbf{E}\{V\}]^{3/2}},$$

where $V := \sigma_R^2(S)$ and $\frac{Var\{V\}}{8\,[\mathbf{E}\{V\}]^{3/2}}$ is the convexity adjustment.

Thus, to calculate volatility swaps, we need both $\mathbf{E}\{V\}$ and $Var\{V\}$.

The realized continuously sampled variance is defined as follows:

$$V := Var(S) := \frac{1}{T}\int_0^T \sigma_t^2 dt.$$

and the realized continuously sampled volatility is defined as

$$Vol(S) := \sqrt{Var(S)} = \sqrt{V}.$$

Options dependent on the exchange rate movements, such as those paying in a currency different from the underlying currency, have an exposure to movements of the correlation between the asset and the exchange rate. This risk may be eliminated by using a covariance swap. Variance and volatility swaps have been studied by Swishchuk (2010). The novelty of the results in this chapter, with respect to Swishchuk (2010), is that we calculate the volatility swap price explicitly. Moreover, we price both a covariance and correlation swap in a two-risky-asset market model.

A *covariance swap* is a covariance forward contract of the underlying rates S^1 and S^2 whose payoff at expiration is equal to

$$N(Cov_R(S^1, S^2) - K_{cov}),$$

where K_{cov} is a strike price, N is the notional amount and $Cov_R(S^1, S^2)$ is a covariance between the two assets S^1 and S^2.

Therefore, a *correlation swap* is a correlation forward contract of two underlying rates S^1 and S^2 whose payoff at expiration is equal to

$$N(Corr_R(S^1, S^2) - K_{corr}),$$

where $Corr(S^1, S^2)$ is a realized correlation of the two underlying assets S^1 and S^2, K_{corr} is a strike price and N is the notional amount.

Pricing covariance swap, from a theoretical point of view, is similar to pricing variance swaps, since

$$Cov_R(S^1, S^2) = \frac{1}{4}\{\sigma_R^2(S^1S^2) - \sigma_R^2(S^1/S^2)\}$$

where S^1 and S^2 are two given assets, $\sigma_R^2(S)$ is a variance swap for underlying assets and $Cov_R(S^1, S^2)$ is a realized covariance of the two underlying assets S^1 and S^2.

Thus, we need to know variances for S^1S^2 and S^1/S^2 (for more detail, see Volume 1, Chapter 4, section 4.2).

Correlation $Corr_R(S^1, S^2)$ is defined as follows:

$$Corr_R(S^1, S^2) = \frac{Cov_R(S^1, S^2)}{\sqrt{\sigma_R^2(S^1)}\sqrt{\sigma_R^2(S^2)}},$$

where $Cov_R(S^1, S^2)$ is defined above and $\sigma_R^2(S^1)$ is defined in section 4.3.1 of Volume 1.

Now, given the two assets S_t^1 and S_t^2, with $t \in [0, T]$, sampled on days $t_0 = 0 < t_1 < t_2 < \ldots < t_n = T$ between today and maturity T, the log-return of each asset is $R_i^j := \log\left(\frac{S_{t_i}^j}{S_{t_{i-1}}^j}\right)$, $i = 1, 2, \ldots, n$, $j = 1, 2$.

Covariance and correlation can be approximated by

$$Cov_n(S^1, S^2) = \frac{n}{(n-1)T} \sum_{i=1}^{n} R_i^1 R_i^2$$

and

$$Corr_n(S^1, S^2) = \frac{Cov_n(S^1, S^2)}{\sqrt{Var_n(S^1)}\sqrt{Var_n(S^2)}},$$

respectively.

The literature devoted to the volatility derivatives is growing. Here, we give a short overview of the latest developments in this area. The non-Gaussian Ornstein–Uhlenbeck stochastic volatility model was used by Benth *et al.* (2007) to study volatility and variance swaps. Broadie and Jain (2008b) evaluated price and hedging strategies for volatility derivatives in the Heston square root stochastic volatility model, and in Broadie and Jain (2008a) they compare results from various models in order to investigate the effect of jumps and discrete sampling on variance and volatility swaps. A pure jump process with independent increment return models were used by Carr *et al.* (2005) to price derivatives written on realized variance, and subsequent developments are shown in Carr and Lee (2007). We also refer to Carr and Lee (2009) for a good survey on volatility derivatives. Da Fonseca *et al.* (2009) analyzed the influence of variance and covariance swap in a market, by solving a portfolio optimization problem in a market with risky assets and volatility derivatives. Correlation swap price has been investigated by Bossu (2005, 2007) for components of an equity index using a statistical method. The paper by Drissien *et al.* (2009) discusses the price of the correlation risk for equity options. Pricing volatility swaps under Heston's model, with regime-switching and pricing options under a generalized Markov-modulated jump-diffusion model, are discussed in Elliott *et al.* (2007b,a), respectively. The paper by Howison *et al.* (2004) considers the pricing of a range of volatility derivatives, including volatility and variance swaps and swaptions. The pricing options on realized variance in the Heston model, with jumps in returns and volatility, are studied in Sepp (2008). An analytical closed-form pricing of pseudo-variance, pseudo-volatility, pseudo-covariance and pseudo-correlation swaps is studied in Swishchuk *et al.* (2002). The paper by Windcliff *et al.* (2006) investigates the behavior and hedging of discretely observed volatility derivatives.

The rest of this chapter is organized as follows. Section 5.2 is devoted to the study of semi-Markov volatilities and their martingale characterization. Section 5.3 deals with the pricing of variance and volatility swaps and numerical evaluations of them for markets with semi-Markov volatilities. Section 5.4 studies the pricing of covariance and correlation swaps for two risky assets in financial markets with semi-Markov volatilities. We also give a numerical evaluation of these derivatives. The Appendix presents a first-order correction for a realized correlation.

5.2. Martingale representation of semi-Markov processes

Let $(\Omega, \mathcal{F}, (\mathcal{F}_t)_{t\in\mathbb{R}_+}, \mathbb{P})$ be a filtered probability space, with a right-continuous filtration $(\mathcal{F}_t)_{t\in\mathbb{R}_+}$ and probability measure $\mathbb{P}$.

Let $(X, \mathcal{X})$ be a measurable space and let

$$Q_{SM}(x, B, t) := P(x, B)G_x(t) \qquad \text{for } x \in X,\ B \in \mathcal{X},\ t \in \mathbb{R}_+, \tag{5.1}$$

be a semi-Markov kernel. Assume that $(x_n, \tau_n; n \in \mathbb{N})$ is a $(X \times \mathbb{R}_+, \mathcal{X} \otimes \mathcal{B}_+)$-valued Markov renewal process with associated kernel Q_{SM}, i.e.

$$\mathbb{P}(x_{n+1} \in B, \tau_{n+1} - \tau_n \leq t \mid \mathcal{F}_n) = Q_{SM}(x_n, B, t). \tag{5.2}$$

Let us define the process

$$\nu_t := \sup\{n \in \mathbb{N} : \tau_n \leq t\} \tag{5.3}$$

that gives the number of jumps of the Markov renewal process in the time interval $(0, t]$, and

$$\theta_n := \tau_n - \tau_{n-1} \tag{5.4}$$

is the sojourn time (or holding time) of the Markov renewal process in the nth visited state. The semi-Markov process, associated with the Markov renewal process $(x_n, \tau_n)_{n\in\mathbb{N}}$, is defined by

$$x_t := x_{\nu(t)}, \qquad \text{for } t \in \mathbb{R}_+. \tag{5.5}$$

It is possible to define some auxiliary processes associated with the semi-Markov process. For instance, we are interested in the backward recurrence time (or life-time) process defined by

$$\gamma(t) := t - \tau_{\nu(t)}, \qquad \text{for } t \in \mathbb{R}_+. \tag{5.6}$$

The next result characterizes the backward recurrence time process (see Swishchuk 2010).

PROPOSITION 5.1.– The backward recurrence time $(\gamma(t))_t$ is a Markov process with generator

$$Q_\gamma f(t) = f'(t) + \lambda(t)[f(0) - f(t)], \tag{5.7}$$

where $\lambda(t) = -\frac{\overline{G}'_x(t)}{\overline{G}_x(t)}$, $\overline{G}_x(t) = 1 - G_x(t)$ and $Domain(Q_\gamma) = C^1(\mathbb{R}_+)$.

DEMONSTRATION.– Let t be the present time, such that $\gamma(t) = t$. Without loss of generality we can assume that $t < \tau_1$, then for $T > t$ we have

$$\mathbf{E}_t\{f(\gamma(T))\} = \mathbf{E}_t\{f(\gamma(T))\mathbb{I}_{\theta_1 > T}\} + \mathbf{E}_t\{f(\gamma(T))\mathbb{I}_{\theta_1 \leq T}\}. \tag{5.8}$$

Using the properties of conditional expectation, we obtain

$$\begin{aligned}\mathbf{E}_t\{f(\gamma(T))\} &= f(T)\frac{\overline{G}_x(T)}{\overline{G}_x(t)} + \frac{1}{\overline{G}_x(t)}\mathbf{E}\{f(\gamma(T))\mathbb{I}_{t<\theta_1\leq T}\}\\ &= f(T)\frac{\overline{G}_x(T)}{\overline{G}_x(t)} + \frac{1}{\overline{G}_x(t)}\int_t^T f(T-u)G'_x(u)du.\end{aligned} \tag{5.9}$$

By adding and subtracting $f(t)$ in the integrand, we get

$$\mathbf{E}_t\{f(\gamma(T))\} = f(T)\frac{\overline{G}_x(T)}{\overline{G}_x(t)} + \frac{1}{\overline{G}_x(t)}\int_t^T (f(T-u) - f(t))G'_x(u)du$$

$$+f(t)\frac{\overline{G}_x(t) - \overline{G}_x(T)}{\overline{G}_x(t)}, \tag{5.10}$$

then

$$\mathbf{E}_t\{f(\gamma(T))\} - f(t) = (f(T) - f(t))\frac{\overline{G}_x(T)}{\overline{G}_x(t)}$$

$$+\frac{1}{\overline{G}_x(t)}\int_t^T (f(T-u) - f(t))G'_x(u)du. \tag{5.11}$$

Now, we recall the definition of the generator, and using the equation above we have

$$Q_\gamma f(t) = \lim_{T\to t}\frac{\mathbf{E}_t\{f(\gamma(T))\} - f(t)}{T-t} = f'(t) - \frac{\overline{G}'_x(t)}{\overline{G}_x(t)}[f(0) - f(t)]. \tag{5.12}$$

This concludes the proof.

REMARK 5.1.– *As it is well known, semi-Markov processes only preserve the memoryless property at transition times, then* $(x_t)_t$ *is not Markov. However, if we consider the joint process* $(x_t, \gamma(t))_{t\in\mathbb{R}_+}$, *then at any instant we record the time already spent by the semi-Markov process in the present state, and it results that* $(x_t, \gamma(t))_{t\in\mathbb{R}_+}$ *is a Markov process.*

The following result allows us to associate a martingale to any Markov process; we refer to Swishchuk (2010) and Elliott and Swishchuk (2007) for details.

PROPOSITION 5.2.– Let $(x_t)_{t\in\mathbb{R}_+}$ be a Markov process with generator Q and $f \in$ Domain (Q), then

$$m_t^f := f(x_t) - f(x_0) - \int_0^t Qf(x_s)ds \qquad [5.13]$$

is a zero-mean martingale with respect to $\mathcal{F}_t := \sigma\{y(s); 0 \le s \le t\}$.

Let us evaluate the quadratic variation of this martingale (see Swishchuk 2010; Elliott and Swishchuk 2007).

PROPOSITION 5.3.– Suppose $(x_t)_{t\in\mathbb{R}_+}$ is a Markov process with generator Q, $f \in$ Domain (Q) and $(m_t^f)_{t\in\mathbb{R}_+}$ is its associated martingale, then

$$\langle m^f\rangle_t := \int_0^t [Qf^2(x_s) - 2f(x_s)Qf(x_s)]ds \qquad [5.14]$$

is the quadratic variation of m^f.

The next result concerns the evaluation of the quadratic covariation; for details and proof, we refer to Salvi and Swishchuk (2012b).

PROPOSITION 5.4.– Let $(x_t)_{t\in\mathbb{R}_+}$ be a Markov process with generator Q, $f, g \in$ Domain (Q) such that $fg \in$ Domain (Q), and denote by $(m_t^f)_{t\in\mathbb{R}_+}, (m_t^g)_{t\in\mathbb{R}_+}$ their associated martingales. Thus,

$$\langle f(x.), g(x.)\rangle_t$$
$$:= \int_0^t \{Q(f(x_s)g(x_s)) - [g(x_s)Qf(x_s) + f(x_s)Qg(x_s)]\}\, ds \qquad [5.15]$$

is the quadratic covariation of f and g.

We would like to study the martingale associated with the Markov process $(x_t, \gamma(t))_{t\in\mathbb{R}_+}$ and its generator. The following statement concern this task, and it is a direct application of proposition 5.2.

LEMMA 5.1 (Swishchuk 2010).– Let $(x_t)_{t\in\mathbb{R}_+}$ be a semi-Markov process with kernel Q_{SM} defined in [5.1]. Then, the process

$$m_t^f := f(x_t, \gamma(t)) - \int_0^t Qf(x_s, \gamma(s))ds \qquad [5.16]$$

is a martingale with respect to the filtration $\mathcal{F}_t := \sigma\{x_s, \nu_s; 0 \leq s \leq t\}$, where Q is the generator of the Markov process $(x_t, \gamma(t))_{t\in\mathbb{R}_+}$ given by

$$Qf(x,t) = \frac{df}{dt}(x,t) + \frac{g_x(t)}{\bar{G}_x(t)} \int_X P(x,dy)[f(y,0) - f(x,t)], \qquad [5.17]$$

here $g_x(t) = \frac{dG_x(t)}{dt}$.

The following statement follows directly from proposition 5.3 and it allows us to evaluate the quadratic variation of the martingale associated with $(x_t, \gamma(t))_{t\in\mathbb{R}_+}$.

LEMMA 5.2.– Let $(x_t)_{t\in\mathbb{R}_+}$ be a semi-Markov process with kernel Q_{SM}, and $(x_t, \gamma(t))_{t\in\mathbb{R}_+}$ is a Markov process with generator Q, $f \in$ Domain (Q), and $(m_t^f)_{t\in\mathbb{R}_+}$ is its associated martingale, then

$$\langle m^f \rangle_t := \int_0^t [Qf^2(x_s, \gamma(s)) - 2f(x_s, \gamma(s))Qf(x_s, \gamma(s))]ds \qquad [5.18]$$

is the quadratic variation of m^f.

5.3. Variance and volatility swaps for financial markets with semi-Markov stochastic volatilities

Let us consider a market model with only two securities, the risk-free bond and the stock. Suppose that the stock price $(S_t)_{t\in\mathbb{R}_+}$ satisfies the following stochastic differential equation:

$$dS_t = S_t(rdt + \sigma(x_t, \gamma(t))dw_t) \qquad [5.19]$$

where w_t is a standard Wiener process independent of (x, γ). We are interested in studying the property of the volatility $\sigma(x, \gamma)$. Salvi and Swishchuk (2012b) have studied properties of volatility modulated by a Markov process, and we will generalize their work to the semi-Markov case. First of all, we study the second moment of the volatility; (see Swishchuk 2010) for details and proof.

PROPOSITION 5.5.– Suppose that $\sigma \in$ Domain (Q). Then,

$$\mathbf{E}\{\sigma^2(x_t, \gamma(t))|\,\mathcal{F}_u\} = \sigma^2(x_u, \gamma(u)) + \int_u^t Q\,\mathbf{E}\{\sigma^2(x_s, \gamma(s))|\,\mathcal{F}_u\}ds \quad [5.20]$$

for any $0 \leq u \leq t$.

REMARK 5.2.– *From proposition 5.5, we can directly solve the equation for* $\mathbf{E}\{\sigma^2(x_t, \gamma(t))|\,\mathcal{F}_u\}$, *and we obtain*

$$\mathbf{E}\{\sigma^2(x_t, \gamma(t))|\,\mathcal{F}_u\} = e^{(t-u)Q}\sigma^2(x_u, \gamma(u)) \quad [5.21]$$

for any $0 \leq u \leq t$.

In this semi-Markov-modulated model, we are able to evaluate high order moments of volatility, e.g. the next result concerns the fourth moment.

PROPOSITION 5.6.– Suppose that $\sigma^2 \in$ Domain (Q).

$$\mathbf{E}\{\sigma^4(x_t, \gamma(t))\} = e^{tQ}\sigma^4(x, 0) \qquad \text{for } t \in \mathbb{R}_+. \quad [5.22]$$

DEMONSTRATION.– If $\sigma^2 \in Domain(Q)$, then from proposition 5.2 we have

$$m_t^{\sigma^2} := \sigma^2(x_t, \gamma(t)) - \sigma^2(x, 0) - \int_0^t Q\sigma^2(x_s, \gamma(s))ds \quad [5.23]$$

as a zero-mean martingale with respect to $\mathcal{F}_t := \sigma\{x_s, \gamma(s); 0 \leq s \leq t\}$, where Q is the infinitesimal generator of the Markov process $(x_t, \gamma(t))_{t\in[0,T]}$ (see lemma 5.2). Then σ^2 satisfies the following stochastic differential equation:

$$d\sigma^2(x_t, \gamma(t)) = Q\sigma^2(x_t, \gamma(t))dt + dm_t^{\sigma^2}. \quad [5.24]$$

By applying the Itô's Lemma, we obtain

$$d\sigma^4(x_t, \gamma(t)) = 2\,\sigma^2(x_t, \gamma(t))\,d\sigma^2(x_t, \gamma(t)) + d\langle m_\cdot^{\sigma^2}\rangle_t. \quad [5.25]$$

We note that the quadratic variation of m^{σ^2} is (see propositon 5.3) given by

$$\langle m^{\sigma^2}\rangle_t := \int_0^t \left[Q\sigma^4(x_s, \gamma(s)) - 2\sigma^2(x_s, \gamma(s))Q\sigma^2(x_s, \gamma(s))\right] ds. \quad [5.26]$$

Substituting the expression for the quadratic variation $\langle m^{\sigma^2}\rangle$ and for σ^2, we obtain that σ^4 satisfies the following stochastic differential equation:

$$d\sigma^4(x_t,\gamma(t)) = Q\sigma^4(x_t,\gamma(t))dt + 2\sigma^2(x_t,\gamma(t))dm_t^{\sigma^2}. \qquad [5.27]$$

By taking the expectation at both sides of this equation, we obtain

$$\mathbf{E}\{\sigma^4(x_t,\gamma(t))\} = \sigma^4(x,0) + \int_0^t Q\mathbf{E}\{\sigma^4(x_s,\gamma(s))\}ds. \qquad [5.28]$$

After solving this differential equation, we finally have

$$\mathbf{E}\{\sigma^4(x_t,\gamma(t))\} = e^{tQ}\,\sigma^4(x,0). \qquad [5.29]$$

It is known that the market model with semi-Markov stochastic volatility is incomplete (see Swishchuk 2010). In order to price the future contracts, we will use the minimal martingale measure and refer to Swishchuk (2010) for the details. Now, we will focus on the evaluation of the price of variance and volatility swaps.

5.3.1. *Pricing of variance swaps*

Let us start from the most straightforward variance swap. Variance swaps are forward contracts on future realized levels of variance. The payoff of a variance swap with expiration date T is given by

$$N(\sigma_R^2(x) - K_{var}) \qquad [5.30]$$

where $\sigma_R^2(x)$ is the realized stock variance over the life of the contract

$$\sigma_R^2(x) := \frac{1}{T}\int_0^T \sigma^2(x_s,\gamma(s))ds, \qquad [5.31]$$

K_{var} is the strike price for variance and N is the notional amount of dollars per annualized variance point. We will assume that $N = 1$ for sake of simplicity. The price of the variance swap is the expected present value of the payoff in the risk-neutral world

$$P_{var}(x) = \mathbf{E}\{e^{-rT}(\sigma_R^2(x) - K_{var})\}. \qquad [5.32]$$

The following result concerns the evaluation of a variance swap in this semi-Markov volatility model. We refer to Swishchuk (2010) for details and proof.

THEOREM 5.1 (Swishchuk 2010).– The present value of a variance swap for semi-Markov stochastic volatility is

$$P_{var}(x) = e^{-rT}\left\{\frac{1}{T}\int_0^T (e^{tQ}\sigma^2(x,0) - K_{var})dt\right\} \qquad [5.33]$$

where Q is the generator of $(x_t, \gamma(t))_t$, that is

$$Qf(x,t) = \frac{df}{dt}(x,t) + \frac{g_x(t)}{\overline{G}_x(t)}\int_X P(x,dy)[f(y,0) - f(x,t)].$$

5.3.1.1. *Example of a variance swap for stochastic volatility driven by a two-state continuous-time Markov chain*

Let Q be a generator of a two-state continuous time Markov chain

$$Q = \begin{pmatrix} q_{11} & q_{12} \\ q_{21} & q_{22} \end{pmatrix}$$

and

$$P(t) = \begin{pmatrix} p_{11}(t) & p_{12}(t) \\ p_{21}(t) & p_{22}(t) \end{pmatrix}$$

be a Markov transition function. Thus,

$$P(t) = e^{tQ}.$$

In this case, the variance takes two values: $\sigma^2(1)$ and $\sigma^2(2)$.

From formula [5.33], it follows that the value of the variance swap in this case is equal to

$$P(i) = e^{-rT}\left(\frac{1}{T}\int_0^T \left[p_{i1}(s)\sigma^2(1) + p_{i2}(s)\sigma^2(2)\right] ds - K_{var}\right) \qquad [5.34]$$

for $i = 1, 2$.

Thus, the value of the variance swap depends on the initial state of the Markov chain.

We should note that if the Markov chain is stationary with ergodic distribution (p_1, p_2), then the value of a variance swap is equal to

$$P = p_1 P(1) + p_2 P(2),$$

where $P(i)$, $i = 1, 2$, are defined in [5.34] (see also Elliott and Swishchuk 2007).

5.3.2. *Pricing of volatility swaps*

Volatility swaps are forward contracts on future realized levels of volatility. The payoff of a volatility swap with maturity T is given by

$$N(\sigma_R(x) - K_{vol}) \tag{5.35}$$

where $\sigma_R(x)$ is the realized stock volatility over the life of the contract

$$\sigma_R(x) := \sqrt{\frac{1}{T}\int_0^T \sigma^2(x_s, \gamma(s))ds}, \tag{5.36}$$

K_{vol} is the strike price for volatility and N is the notional amount of dollars per annualized volatility point. As before, we will assume that $N = 1$. The price of the volatility swap is the expected present value of the payoff in the risk-neutral world, i.e.

$$P_{vol}(x) = \mathbf{E}\{e^{-rT}(\sigma_R(x) - K_{vol})\}. \tag{5.37}$$

In order to evaluate volatility swaps, we need to know the expected value of the square root of the variance, but unfortunately, in general we are not able to analytically evaluate this expected value. Then, in order to obtain a closed-form formula for the price of volatility swaps, we have to make an approximation. Using the same approach of the Markov case (see also Brockhaus and Long 2000; Javaheri *et al.* 2002), from the second-order Taylor expansion, we have

$$\mathbf{E}\left\{\sqrt{\sigma_R^2(x)}\right\} \approx \sqrt{\mathbf{E}\{\sigma_R^2(x)\}} - \frac{Var\{\sigma_R^2(x)\}}{8\,[\mathbf{E}\{\sigma_R^2(x)\}]^{3/2}}. \tag{5.38}$$

Then, to evaluate the volatility swap price we have to know both the expectation and variance of $\sigma_R^2(x)$. The next result gives an explicit representation of the price of a volatility swap approximated to the second order for this semi-Markov volatility model.

THEOREM 5.2.– The value of a volatility swap for semi-Markov stochastic volatility is

$$P_{vol}(x) \approx e^{-rT}$$
$$\times \left\{ \sqrt{\frac{1}{T}\int_0^T e^{tQ}\sigma^2(x,0)dt} - \frac{Var\{\sigma_R^2(x)\}}{8\left(\frac{1}{T}\int_0^T e^{tQ}\sigma^2(x,0)dt\right)^{3/2}} - K_{vol} \right\}$$

where the variance is given by

$$Var\{\sigma_R^2(x)\} = \frac{2}{T^2}\int_0^T\int_0^t \Big\{e^{sQ}\left[\sigma^2(x,0)e^{(t-s)Q}\sigma^2(x,0)\right]$$

$$-\left[e^{tQ}\sigma^2(x,0)\right]\left[e^{sQ}\sigma^2(x,0)\right]\Big\}dsdt, \qquad [5.39]$$

and Q is the generator of $(x_t, \gamma(t))_t$, that is

$$Qf(x,t) = \frac{df}{dt}(x,t) + \frac{g_x(t)}{\bar{G}_x(t)}\int_X P(x,dy)[f(y,0) - f(x,t)]. \qquad [5.40]$$

DEMONSTRATION.– We have already obtained the expectation of the realized variance,

$$\mathbf{E}\{\sigma_R^2(x)\} = \frac{1}{T}\int_0^T e^{tQ}\sigma^2(x,0)dt \qquad [5.41]$$

then it remains to prove that

$$Var\{\sigma_R^2(x)\} = \frac{2}{T^2}\int_0^T\int_0^t \Big\{e^{sQ}\left[\sigma^2(x,0)e^{tQ}\sigma^2(x,0)\right]$$

$$-\left[e^{tQ}\sigma^2(x,0)\right]\left[e^{sQ}\sigma^2(x,0)\right]\Big\}dsdt. \qquad [5.42]$$

The variance is, from the definition, given by

$$Var\{\sigma_R^2(x)\} = \mathbf{E}\{[\sigma_R^2(x) - \mathbf{E}\{\sigma_R^2(x)\}]^2\}. \qquad [5.43]$$

Using the definition of realized variance, and the Fubini theorem, we have

$$Var\{\sigma_R^2(x)\} = \mathbf{E}\left\{\left[\frac{1}{T}\int_0^T \sigma^2(x_t,\gamma(t))dt - \frac{1}{T}\int_0^T \mathbf{E}\{\sigma^2(x_t,\gamma(t))\}dt\right]^2\right\}$$

$$= \mathbf{E}\left\{\left[\frac{1}{T}\int_0^T \left(\sigma^2(x_t,\gamma(t)) - \mathbf{E}\{\sigma^2(x_t,\gamma(t))\}\right)dt\right]^2\right\}$$

$$= \frac{1}{T^2}\int_0^T\int_0^T \mathbf{E}\left\{[\sigma^2(x_t,\gamma(t)) - \mathbf{E}\{\sigma^2(x_t,\gamma(t))\}]\right.$$

$$\left.\times\ [\sigma^2(x_s,\gamma(s)) - \mathbf{E}\{\sigma^2(x_s,\gamma(s))\}]\right\} dsdt.$$

Note that the integrand is symmetric in the exchange of s and t. We can divide the integration on the plane, in two areas above and below the graph of $t = s$; due to the symmetry, the contribution on the two parts is the same. Then, we obtain:

$$Var\{\sigma_R^2(x)\} = \frac{2}{T^2}\int_0^T\int_0^t \mathbf{E}\{[\sigma^2(x_t,\gamma(t)) - \mathbf{E}\{\sigma^2(x_t,\gamma(t))\}]$$
$$\times[\sigma^2(x_s,\gamma(s)) - \mathbf{E}\{\sigma^2(x_s,\gamma(s))\}]\}dsdt$$
$$= \frac{2}{T^2}\int_0^T\int_0^t \left[\mathbf{E}\{\sigma^2(x_t,\gamma(t))\sigma^2(x_s,\gamma(s))\}\right.$$
$$\left. - \mathbf{E}\{\sigma^2(x_t,\gamma(t))\}\mathbf{E}\{\sigma^2(x_s,\gamma(s))\}\right] dsdt.$$

We would like to stress that, in this representation, the integration set is such that the inequality $s \leq t$ holds true. Using the property of conditional expectation and proposition 5.5, we have

$$Var\{\sigma_R^2(x)\} = \frac{2}{T^2}\int_0^T\int_0^t \left[\mathbf{E}\{\sigma^2(x_s,\gamma(s))\mathbf{E}\{\sigma^2(x_t,\gamma(t))|\mathcal{F}_s\}\}\right.$$
$$\left. - \left(e^{tQ}\sigma^2(x,0)\right)\left(e^{sQ}\sigma^2(x,0)\right)\right] dsdt.$$

The process $(x_t,\gamma(t))_t$ is Markov, then using remark 5.2 the conditional expected value in the integrand can be expressed as

$$\mathbf{E}\{\sigma^2(x_t,\gamma(t))|\mathcal{F}_s\} = e^{(t-s)Q}\sigma^2(x_s,\gamma(s)) =: g(x_s,\gamma(s)). \qquad [5.44]$$

Thus, the variance becomes

$$Var\{\sigma_R^2(x)\} = \frac{2}{T^2}\int_0^T\int_0^t \left[\mathbf{E}\{\sigma^2(x_s,\gamma(s))g(x_s,\gamma(s))\}\right.$$
$$\left. - \left(e^{tQ}\sigma^2(x,0)\right)\left(e^{sQ}\sigma^2(x,0)\right)\right] dsdt.$$

Solving the expectation on the right-hand side, we obtain

$$Var\{\sigma_R^2(x)\} = \frac{2}{T^2}\int_0^T\int_0^t \left[e^{sQ}\left(\sigma^2(x,0)g(x,0)\right)\right.$$
$$\left. - \left(e^{tQ}\sigma^2(x,0)\right)\left(e^{sQ}\sigma^2(x,0)\right)\right] dsdt.$$

We note that function g evaluated in $(x,0)$ is simply given by

$$g(x,0) = e^{(t-s)Q}\sigma^2(x,0), \qquad [5.45]$$

and by substituting in the previous formula, the variance finally becomes

$$Var\{\sigma_R^2(x)\} = \frac{2}{T^2}\int_0^T\int_0^t \left\{e^{sQ}\left[\sigma^2(x,0)e^{(t-s)Q}\sigma^2(x,0)\right]\right.$$
$$\left. - \left[e^{tQ}\sigma^2(x,0)\right]\left[e^{sQ}\sigma^2(x,0)\right]\right\} dsdt.$$

5.3.3. *Numerical evaluation of variance and volatility swaps with semi-Markov volatility*

In the application where we attempt to evaluate the price of a variance or a volatility swap, we have to deal with numerical problems. The family of exponential operators $(e^{tQ})_t$ involved in both theorems 5.1 and 5.2 for the semi-Markov stochastic volatility model, is usually difficult to evaluate from a numerical point of view. To solve this problem, we first look to the following identity:

$$e^{tQ}f(\cdot) = \sum_{n=0}^{\infty} \frac{(tQ)^n}{n!} f(\cdot), \qquad [5.46]$$

for any function $f \in Domain(Q)$. This identity allow us to obtain the operator $(e^{tQ})_t$ at any order of approximation. For example, for $n = 1$, we obtain

$$e^{tQ}f(\cdot) \approx (I + tQ)f(\cdot) \qquad [5.47]$$

where I is an identity operator. At this order of approximation, we allow the semi-Markov process to make, at most, one transition during the life time of the contract. If we think of a semi-Markov process as a macroeconomic factor, this can be plausible. Moreover, we can always evaluate the error in this approximation using the subsequent orders. Using the first-order approximation, the variance swap price becomes

$$\begin{aligned} P_{var}(x) &\approx e^{-rT}\left\{\frac{1}{T}\int_0^T (I + tQ)\sigma^2(x,0)dt - K_{var}\right\} \\ &= e^{-rT}\left\{\sigma^2(x,0) + \frac{T}{2}Q\sigma^2(x,0) - K_{var}\right\}. \end{aligned} \qquad [5.48]$$

Using the same approximation, the volatility swap price can be expressed as

$$\begin{aligned} P_{vol}(x) &\approx e^{-rT}\Bigg\{\sqrt{\frac{1}{T}\int_0^T (I + tQ)\sigma^2(x,0)dt} \\ &\quad - \frac{Var\{\sigma_R^2(x)\}}{8\left(\frac{1}{T}\int_0^T (I + tQ)\sigma^2(x,0)dt\right)^{3/2}} - K_{vol}\Bigg\} \\ &= e^{-rT}\Bigg\{\sqrt{\sigma^2(x,0) + \frac{T}{2}Q\sigma^2(x,0)} \end{aligned}$$

$$
- \frac{Var\{\sigma_R^2(x)\}}{8\left(\sigma^2(x,0) + \frac{T}{2}Q\sigma^2(x,0)\right)^{3/2}} - K_{vol} \Bigg\}. \quad [5.49]
$$

Here, the variance of realized volatility is given by

$$
\begin{aligned}
Var\{\sigma_R^2(x)\} \approx & \frac{2}{T^2} \int_0^T \int_0^t \{(I + sQ)[\sigma^2(x,0)(I + (t-s)Q)\sigma^2(x,0)] \\
& - [(I + tQ)\sigma^2(x,0)][(I + sQ)\sigma^2(x,0)]\} \, dsdt.
\end{aligned} \quad [5.50]
$$

Solving the product and by only keeping the terms up to the first order in Q, we obtain

$$
\begin{aligned}
Var\{\sigma_R^2(x)\} \approx & \frac{2}{T^2} \int_0^T \int_0^t \{sQ\sigma^4(x,0) - 2\sigma^2(x,0)sQ\sigma^2(x,0)\} dsdt \\
= & \frac{T}{3} \left\{Q\sigma^4(x,0) - 2\sigma^2(x,0)Q\sigma^2(x,0)\right\}.
\end{aligned} \quad [5.51]
$$

Finally, at the first order of approximation in Q, the volatility swap price becomes

$$
\begin{aligned}
P_{vol}(x) \approx e^{-rT} \Bigg\{ & \sqrt{\sigma^2(x,0) + \frac{T}{2}Q\sigma^2(x,0)} \\
& - \frac{T[Q\sigma^4(x,0)2\sigma^2(x,0)Q\sigma^2(x,0)]}{24\left(\sigma^2(x,0) + \frac{T}{2}Q\sigma^2(x,0)\right)^{3/2}} - K_{vol} \Bigg\}.
\end{aligned} \quad [5.52]
$$

5.4. Covariance and correlation swaps for two risky assets in financial markets with semi-Markov stochastic volatilities

Let us now consider a market model with two risky assets and one risk-free bond. Let us assume that the risky assets are satisfying the following stochastic differential equations:

$$
\begin{cases}
dS_t^{(1)} = S_t^{(1)}(\mu_t^{(1)}dt + \sigma^{(1)}(x_t, \gamma(t))dw_t^{(1)}) \\
dS_t^{(2)} = S_t^{(2)}(\mu_t^{(2)}dt + \sigma^{(2)}(x_t, \gamma(t))dw_t^{(2)})
\end{cases} \quad [5.53]
$$

where $\mu^{(1)}$, $\mu^{(2)}$ are deterministic functions of time, and $(w_t^{(1)})_t$ and $(w_t^{(2)})_t$ are standard Wiener processes with quadratic covariance, given by

$$
d[w_t^{(1)}, w_t^{(2)}] = \rho_t \, dt, \quad [5.54]
$$

where ρ_t is a deterministic function and $(w_t^{(1)})_t, (w_t^{(2)})_t$ are independent of (x, γ).

In this model, it is worth studying the covariance and the correlation swaps between the two risky assets.

5.4.1. *Pricing of covariance swaps*

A covariance swap is a covariance forward contract on the underlying assets $S^{(1)}$ and $S^{(2)}$ whose payoff at maturity is equal to

$$N(Cov_R(S^{(1)}, S^{(2)}) - K_{cov}) \tag{5.55}$$

where K_{cov} is a strike reference value, N is the notional amount and the realized covariance of the two assets $S^{(1)}$ and $S^{(2)}$, say $Cov_R(S^{(1)}, S^{(2)})$, is given by

$$Cov_R(S^{(1)}, S^{(2)}) = \frac{1}{T}\left[\ln S_T^{(1)}, \ln S_T^{(2)}\right]$$

$$= \frac{1}{T}\int_0^T \rho_t\, \sigma^{(1)}(x_t, \gamma(t))\sigma^{(2)}(x_t, \gamma(t))dt. \tag{5.56}$$

The price of the covariance swap is the expected present value of the payoff in the risk-neutral world, i.e.

$$P_{cov}(x) = \mathbf{E}\{e^{-rT}(Cov_R(S^{(1)}, S^{(2)}) - K_{cov})\}, \tag{5.57}$$

where we assume $N = 1$. The next result provides us with an explicit representation of the covariance swap price.

THEOREM 5.3.– The value of a covariance swap for semi-Markov stochastic volatility is

$$P_{cov}(x) = e^{-rT}\left\{\frac{1}{T}\int_0^T \rho_t e^{tQ}[\sigma^{(1)}(x,0)\sigma^{(2)}(x,0)]dt - K_{cov}\right\}, \tag{5.58}$$

where Q is the generator of $(x_t, \gamma(t))_t$, that is

$$Qf(x,t) = \frac{df}{dt}(x,t) + \frac{g_x(t)}{\overline{G}_x(t)}\int_X P(x,dy)[f(y,0) - f(x,t)]. \tag{5.59}$$

DEMONSTRATION.– To evaluate the price of covariance swap, we need to know

$$\mathbf{E}\{Cov_R(S^{(1)}, S^{(2)})\} = \frac{1}{T}\int_0^T \rho_t \mathbf{E}\{\sigma^{(1)}(x_t, \gamma(t))\sigma^{(2)}(x_t, \gamma(t))\}dt. \quad [5.60]$$

It remains to prove that

$$\mathbf{E}\{\sigma^{(1)}(x_t, \gamma(t))\sigma^{(2)}(x_t, \gamma(t))\} = e^{tQ}[\sigma^{(1)}(x, 0)\sigma^{(2)}(x, 0)]. \quad [5.61]$$

By applying Itô's lemma, we have

$$d(\sigma^{(1)}(x_t, \gamma(t))\sigma^{(2)}(x_t, \gamma(t))) = \sigma^{(1)}(x_t, \gamma(t))d\sigma^{(2)}(x_t, \gamma(t))$$
$$+\sigma^{(2)}(x_t, \gamma(t))d\sigma^{(1)}(x_t, \gamma(t)) + d\langle\sigma^{(1)}(x_\cdot, \gamma(\cdot)), \sigma^{(2)}(x_\cdot, \gamma(\cdot))\rangle_t. \quad [5.62]$$

Using proposition 5.4, we obtain

$$d\langle\sigma^{(1)}(x_\cdot, \gamma(\cdot)), \sigma^{(2)}(x_\cdot, \gamma(\cdot))\rangle_t = Q(\sigma^{(1)}(x_t, \gamma(t))\sigma^{(2)}(x_t, \gamma(t)))dt$$
$$-[\sigma^{(1)}(x_t, \gamma(t))Q\sigma^{(2)}(x_t, \gamma(t)) + \sigma^{(2)}(x_t, \gamma(t))Q\sigma^{(1)}(x_t, \gamma(t))]dt. \quad [5.63]$$

Furthermore, we have

$$d\sigma^{(i)}(x_t, \gamma(t)) = Q\sigma^{(i)}(x_t, \gamma(t))dt + dm^{\sigma^{(i)}}, \qquad i = 1, 2. \quad [5.64]$$

By substituting [5.63] and [5.64] into equation [5.62], we obtain

$$d(\sigma^{(1)}(x_t, \gamma(t))\sigma^{(2)}(x_t, \gamma(t))) = Q\,(\sigma^{(1)}(x_t, \gamma(t))\sigma^{(2)}(x_t, \gamma(t)))dt$$
$$+\sigma^{(1)}(x_t, \gamma(t))dm^{\sigma^{(2)}} + \sigma^{(2)}(x_t, \gamma(t))dm^{\sigma^{(1)}}. \quad [5.65]$$

Taking the expectation at both sides, we can rewrite the above equation as

$$\mathbf{E}\{\sigma^{(1)}(x_t, \gamma(t))\sigma^{(2)}(x_t, \gamma(t))\} = \sigma^{(1)}(x, 0)\sigma^{(2)}(x, 0)$$
$$+\int_0^t Q\,\mathbf{E}\{\sigma^{(1)}(x_s, \gamma(s))\sigma^{(2)}(x_s, \gamma(s))\}dt. \quad [5.66]$$

and solving this differential equation gives

$$\mathbf{E}\{\sigma^{(1)}(x_t, \gamma(t))\sigma^{(2)}(x_t, \gamma(t))\} = e^{tQ}[\sigma^{(1)}(x, 0)\sigma^{(2)}(x, 0)]. \quad [5.67]$$

This concludes the proof.

5.4.2. *Pricing of correlation swaps*

A correlation swap is a forward contract on the correlation between the underlying assets S^1 and S^2 whose payoff at maturity is equal to

$$N(Corr_R(S^1, S^2) - K_{corr}) \quad [5.68]$$

where K_{corr} is a strike reference level, N is the notional amount and the realized correlation, say $Corr_R(S^1, S^2)$, is defined by

$$Corr_R(S^1, S^2) = \frac{Cov_R(S^1, S^2)}{\sqrt{\sigma_R^{(1)^2}(x)}\sqrt{\sigma_R^{(2)^2}(x)}}, \quad [5.69]$$

where the realized variance is given by

$$\sigma_R^{(i)^2}(x) = \frac{1}{T}\int_0^T (\sigma^{(i)}(x_t, \gamma(t)))^2 dt, \qquad i = 1, 2. \quad [5.70]$$

The price of the correlation swap is the expected present value of the payoff in the risk-neutral world, that is

$$P_{corr}(x) = \mathbf{E}\{e^{-rT}(Corr_R(S^1, S^2) - K_{corr})\} \quad [5.71]$$

where we set $N = 1$ for simplicity. Unfortunately, the expected value of $Corr_R(S^1, S^2)$ is not known analytically. Thus, in order to obtain an explicit formula for the correlation swap price, we have to make some approximations.

First of all, let us introduce the following notations:

$$\begin{aligned} X &= Cov_R(S^1, S^2) \\ Y &= \sigma_R^{(1)^2}(x) \\ Z &= \sigma_R^{(2)^2}(x), \end{aligned} \quad [5.72]$$

and with the subscript 0, we will denote the expected value of the above random variables. Following the approach frequently used for the volatility swap, we would like to approximate the square root of Y and Z at the first order as follows:

$$\begin{aligned} \sqrt{Y} &\approx \sqrt{Y_0} + \frac{Y - Y_0}{2\sqrt{Y_0}} \\ \sqrt{Z} &\approx \sqrt{Z_0} + \frac{Z - Z_0}{2\sqrt{Z_0}}. \end{aligned} \quad [5.73]$$

The realized correlation can now be approximated by

$$\begin{aligned} Corr_R(S^1, S^2) &\approx \frac{X}{\left(\sqrt{Y_0} + \frac{Y-Y_0}{2\sqrt{Y_0}}\right)\left(\sqrt{Z_0} + \frac{Z-Z_0}{2\sqrt{Z_0}}\right)} \\ &= \frac{\frac{X}{\sqrt{Y_0}\sqrt{Z_0}}}{\left(1 + \frac{Y-Y_0}{2Y_0}\right)\left(1 + \frac{Z-Z_0}{2Z_0}\right)}. \end{aligned} \quad [5.74]$$

Solving the product in the denominator on the last term of the right-hand side, and by only keeping the terms up to the first order in the increment, we have

$$\begin{aligned} Corr_R(S^1, S^2) &\approx \frac{\frac{X}{\sqrt{Y_0}\sqrt{Z_0}}}{1 + \left(\frac{Y-Y_0}{2Y_0} + \frac{Z-Z_0}{2Z_0}\right)} \\ &\approx \frac{X}{\sqrt{Y_0}\sqrt{Z_0}}\left[1 - \left(\frac{Y - Y_0}{2Y_0} + \frac{Z - Z_0}{2Z_0}\right)\right]. \end{aligned} \quad [5.75]$$

In the following, we will only consider the zero order of approximation, and we will discuss the first correction in the Appendix. Here, we are going to approximate the realized correlation as

$$Corr_R(S^1, S^2) \approx \frac{X}{\sqrt{Y_0}\sqrt{Z_0}} \quad [5.76]$$

Substituting X, Y and Z we obtain

$$\begin{aligned} Corr_R(S^1, S^2) &\approx \frac{1}{\sqrt{\mathbf{E}\{\sigma_R^{(1)^2}(x)\}}\sqrt{\mathbf{E}\{\sigma_{2R}^{(2)^2}(x)\}}} \\ &\times \frac{1}{T}\int_0^T \rho_t\, \sigma^{(1)}(x_t, \gamma(t))\sigma^{(2)}(x_t, \gamma(t))dt \end{aligned} \quad [5.77]$$

where (see theorem 5.1), we have

$$\mathbf{E}\left\{\sigma_R^{(i)^2}(x)\right\} = \mathbf{E}\left\{\frac{1}{T}\int_0^T (\sigma^{(i)}(x_t, \gamma(t)))^2 dt\right\}$$

$$= \frac{1}{T}\int_0^T e^{tQ}(\sigma^{(i)}(x,0))^2 dt, \qquad [5.78]$$

for $i = 1, 2$. In order to price a correlation swap, we need to evaluate the expectation at both sides of the equation [5.77]. The expectation of the right-hand side becomes

$$\mathbf{E}\left\{\sigma^{(1)}(x_t, \gamma(t))\sigma^{(2)}(x_t, \gamma(t))\right\} = e^{tQ}\sigma^{(1)}(x,0)\sigma^{(2)}(x,0). \qquad [5.79]$$

We can summarize the previous result in the following statement.

THEOREM 5.4.– The value of a correlation swap for semi-Markov stochastic volatility is

$$P_{corr}(x) \approx e^{-rT}$$

$$\times \left\{\frac{\int_0^T \rho_t e^{tQ}[\sigma^{(1)}(x,0)\sigma^{(2)}(x,0)]dt}{\sqrt{\int_0^T e^{tQ}(\sigma^{(1)}(x,0))^2 dt}\sqrt{\int_0^T e^{tQ}(\sigma^{(2)}(x,0))^2 dt}} - K_{corr}\right\}, \qquad [5.80]$$

where Q is the generator of $(x_t, \gamma(t))_t$, i.e.

$$Qf(x,t) = \frac{df}{dt}(x,t) + \frac{g_x(t)}{G_x(t)}\int_X P(x,dy)[f(y,0) - f(x,t)]. \qquad [5.81]$$

5.5. Numerical evaluation of covariance and correlation swaps with semi-Markov stochastic volatility

In order to obtain a more handy expression for the price of covariance and correlation swaps to use in the application, we will introduce an approximation for the family of operator $(e^{tQ})_t$. Following a similar approach used for the variance and volatility cases, we are going to approximate the operators at the first order in Q as

$$e^{tQ}f(\cdot) \approx (I + tQ)f(\cdot). \qquad [5.82]$$

Using this approximation, the covariance swap price becomes

$$P_{cov}(x) \approx e^{-rT}\left\{\frac{1}{T}\int_0^T \rho_t(I + tQ)[\sigma^{(1)}(x,0)\sigma^{(2)}(x,0)]dt - K_{cov}\right\}$$

$$= e^{-rT}\Big\{\sigma^{(1)}(x,0)\sigma^{(2)}(x,0)\int_0^T \rho_t dt + Q[\sigma^{(1)}(x,0)\sigma^{(2)}(x,0)] \times \int_0^T t\rho_t dt - K_{cov}\Big\} \quad [5.83]$$

The same approximation allows us to express the correlation swap price as

$$P_{corr}(x) \approx e^{-rT} \times \left\{ \frac{\int_0^T \rho_t (I+tQ)[\sigma^{(1)}(x,0)\sigma^{(2)}(x,0)]dt}{\sqrt{\int_0^T (I+tQ)(\sigma^{(1)}(x,0))^2 dt}\sqrt{\int_0^T (I+tQ)(\sigma^{(2)}(x,0))^2 dt}} - K_{corr} \right\}$$

$$= e^{-rT} \times \left\{ \frac{\sigma^{(1)}(x,0)\sigma^{(2)}(x,0)\int_0^T \rho_t dt + Q[\sigma^{(1)}(x,0)\sigma^{(2)}(x,0)]\int_0^T t\rho_t dt}{\sqrt{(\sigma^{(1)}(x,0))^2 + \frac{T}{2}Q(\sigma^{(1)}(x,0))^2}\sqrt{(\sigma^{(2)}(x,0))^2 + \frac{T}{2}Q(\sigma^{(2)}(x,0))^2}} - K_{corr} \right\}. \quad [5.84]$$

5.6. Appendices

5.6.1. *Appendix 1. Realized correlation: first-order correction*

We would like to obtain an approximation for the realized correlation between two risky assets

$$Corr_R(S^1, S^2) = \frac{Cov_R(S^1, S^2)}{\sqrt{\sigma_R^{(1)^2}(x)}\sqrt{\sigma_R^{(2)^2}(x)}}. \quad [5.85]$$

In section 5.4.2, we obtained the following approximated expression:

$$Corr_R(S^1, S^2) \approx \frac{\frac{X}{\sqrt{Y_0}\sqrt{Z_0}}}{1 + \left(\frac{Y-Y_0}{2Y_0} + \frac{Z-Z_0}{2Z_0}\right)} \approx \frac{X}{\sqrt{Y_0}\sqrt{Z_0}}\left[1 - \left(\frac{Y-Y_0}{2Y_0} + \frac{Z-Z_0}{2Z_0}\right)\right], \quad [5.86]$$

where

$$\begin{aligned} X &= Cov_R(S^1, S^2) \\ Y &= \sigma_R^{(1)^2}(x) \\ Z &= \sigma_R^{(2)^2}(x), \end{aligned} \tag{5.87}$$

and with the subscript 0, we have denoted the expected values. We have already evaluated the expectation of the zero-order approximation; now, we would like to evaluate the first-order approximation.

Substituting X, Y and Z into equation [5.86], we obtain

$$\begin{aligned} Corr_R(S^1, S^2) \approx & \frac{1}{\sqrt{\mathbf{E}\{\sigma_R^{(1)^2}(x)\}}\sqrt{\mathbf{E}\{\sigma_{2R}^{(2)^2}(x)\}}} \\ & \times \frac{1}{T}\int_0^T \rho_t \sigma^{(1)}(x_t, \gamma(t))\sigma^{(2)}(x_t, \gamma(t))dt \\ & - \frac{1}{2T^2(\mathbf{E}\{\sigma_R^{(1)^2}(x)\})^{3/2}(\mathbf{E}\{\sigma_R^{(2)^2}(x)\})^{3/2}} \int_0^T \rho_t \sigma^{(1)}(x_t, \gamma(t))\sigma^{(2)}(x_t, \gamma(t))dt \\ & \times \Bigg\{ \mathbf{E}\{\sigma_R^{(2)^2}(x)\} \int_0^T [(\sigma^{(1)}(x_s, \gamma(s)))^2 - \mathbf{E}\{(\sigma^{(1)}(x_s, \gamma(s)))^2\}]ds \\ & + \mathbf{E}\{\sigma_R^{(1)^2}(x)\} \int_0^T [(\sigma^{(2)}(x_u, \gamma(u)))^2 - \mathbf{E}\{(\sigma^{(2)}(x_u, \gamma(u)))^2\}]du \Bigg\}, \end{aligned} \tag{5.88}$$

where

$$\begin{aligned} \mathbf{E}\left\{\sigma_{(i)R}^2(x)\right\} &= \mathbf{E}\left\{\frac{1}{T}\int_0^T (\sigma^{(i)}(x_t, \gamma(t)))^2 dt\right\} \\ &= \frac{1}{T}\int_0^T e^{tQ}(\sigma^{(i)}(x, 0))^2 dt, \end{aligned} \tag{5.89}$$

for $i = 1, 2$. We have to evaluate the expectation of the right-hand side of equation [5.88].We have already calculated the expectation of the first term, which is the zero-order approximation for the realized correlation. We now focus on the other terms. First of all, let us rewrite them as follows:

$$\begin{aligned}\int_0^T\int_0^T & \rho_t\sigma^{(1)}(x_t,\gamma(t))\sigma^{(2)}(x_t,\gamma(t))\left(\mathbf{E}\{\sigma_R^{(2)^2}(x)\}[(\sigma^{(1)}(x_s,\gamma(s)))^2\right.\\ & -\mathbf{E}\{(\sigma^{(1)}(x_s,\gamma(s)))^2\}]\\ & \left.+\mathbf{E}\{\sigma_R^{(1)^2}(x)\}[(\sigma^{(2)}(x_s,\gamma(s)))^2-\mathbf{E}\{(\sigma^{(2)}(x_s,\gamma(s)))^2\}]\right)dsdt.\end{aligned} \quad [5.90]$$

We have four different contributions in the integrals; the expectation of the terms

$$\begin{aligned}\int_0^T\int_0^T & \rho_t\sigma^{(1)}(x_t,\gamma(t))\sigma^{(2)}(x_t,\gamma(t))\\ & \times\mathbf{E}\{\sigma^{(i)^2}(x_s,\gamma(s))\}\mathbf{E}\{\sigma^{(-i)^2}(x_s,\gamma(s))\}dsdt\end{aligned} \quad [5.91]$$

for $i = 1,2$, can be evaluated using theorem 5.3. Then, in order to evaluate the expectation of the approximated realized correlation, it only remains to calculate, for $i = 1,2$

$$\mathbf{E}\left\{\int_0^T\int_0^T\rho_t\sigma^{(1)}(x_t,\gamma(t))\sigma^{(2)}(x_t,\gamma(t))\sigma^{(i)^2}(x_s,\gamma(s))dsdt\right\}. \quad [5.92]$$

To this end, let us first divide the range of integration in two intervals as follows:

$$\begin{aligned}\mathbf{E}\Bigg\{ & \int_0^T\int_0^t\rho_t\sigma^{(1)}(x_t,\gamma(t))\sigma^{(2)}(x_t,\gamma(t))\sigma^{(i)^2}(x_s,\gamma(s))dsdt\\ & +\int_0^T\int_t^T\rho_t\sigma^{(1)}(x_t,\gamma(t))\sigma^{(2)}(x_t,\gamma(t))\sigma^{(i)^2}(x_s,\gamma(s))dsdt\Bigg\}\end{aligned} \quad [5.93]$$

for $i = 1,2$. We note that the first integral set is such that $t > s$, whereas the second one has $t < s$. We can now use the property of conditional expectation to obtain

$$\begin{aligned}\mathbf{E}\Bigg\{ & \int_0^T\int_0^t\rho_t\mathbf{E}\{\sigma^{(1)}(x_t,\gamma(t))\sigma^{(2)}(x_t,\gamma(t))|\mathcal{F}_s\}\sigma^{(i)^2}(x_s,\gamma(s))dsdt\\ & +\int_0^T\int_t^T\rho_t\sigma^{(1)}(x_t,\gamma(t))\sigma^{(2)}(x_t,\gamma(t))\mathbf{E}\{\sigma^{(i)^2}(x_s,\gamma(s))|\mathcal{F}_t\}dsdt\Bigg\}.\end{aligned} \quad [5.94]$$

We should retain that $(x_t, \gamma(t))_t$ is a Markov process; then using the Markov property, we can express the conditional expectations as

$$\begin{aligned}\mathbf{E}\{\sigma^{(1)}(x_t,\gamma(t))\sigma^{(2)}(x_t,\gamma(t))|\,\mathcal{F}_s\} &= e^{(t-s)Q}\sigma^{(1)}(x_s,\gamma(s))\sigma^{(2)}(x_s,\gamma(s)) \\ &=: h(x_s,\gamma(s))\end{aligned} \quad [5.95]$$

for $t > s$, and

$$\mathbf{E}\{\sigma^{(i)^2}(x_s,\gamma(s))|\,\mathcal{F}_t\} = e^{(s-t)Q}\sigma^{(i)^2}(x_t,\gamma(t)) =: g^{(i)}(x_t,\gamma(t))$$

for $s > t$. Therefore, the first term of equation [5.94] can be expressed as

$$\begin{aligned}&\mathbf{E}\left\{\int_0^T\int_0^t \rho_t h(x_s,\gamma(s))\sigma^{(i)^2}(x_s,\gamma(s))dsdt\right\} \\ &= \int_0^T\int_0^t \rho_t e^{sQ}[h(x,0)\sigma^{(i)^2}(x,0)]dsdt,\end{aligned} \quad [5.96]$$

whereas the second term can be expressed as

$$\begin{aligned}&\mathbf{E}\left\{\int_0^T\int_t^T \rho_t\sigma^{(1)}(x_t,\gamma(t))\sigma^{(2)}(x_t,\gamma(t))g^{(i)}(x_t,\gamma(t))dsdt\right\} \\ &= \int_0^T\int_t^T \rho_t e^{tQ}[\sigma^{(1)}(x,0)\sigma^{(2)}(x,0)g^{(i)}(x,0)]dsdt.\end{aligned} \quad [5.97]$$

Now, we can evaluate the functions h and g at x obtaining

$$h(x,0) = e^{(t-s)Q}[\sigma^{(1)}(x,0)\sigma^{(2)}(x,0)] \quad [5.98]$$

and

$$g^{(i)}(x,0) = e^{(s-t)Q}[\sigma^{(i)^2}(x,0)]. \quad [5.99]$$

We can summarize the previous result in the following statement that gives the correlation swap price up to the first order of approximation.

THEOREM 5.5.– The value of the correlation swap for a semi-Markov volatility is given by

$$P_{corr}(x) = e^{-rT}\left(\mathbf{E}\{Corr_R(S^1, S^2)\} - K_{corr}\right) \qquad [5.100]$$

where the realized correlation can be approximated by

$$\mathbf{E}\{Corr_R(S^1, S^2)\} \approx \frac{2\int_0^T \rho_t e^{tQ}\sigma^{(1)}(x,0)\sigma^{(2)}(x,0)dt}{\sqrt{K_{10}}\sqrt{K_{20}}}$$

$$-\frac{1}{2K_{10}^{3/2}K_{20}^{1/2}}\Big\{\int_0^T \rho_t\Big(\int_0^t e^{sQ}\left\{e^{tQ}[\sigma^{(1)}(x,0)\sigma^{(2)}(x,0)]\sigma^{(1)^2}(x,0)\right\}ds$$

$$+\int_t^T e^{tQ}\left\{\sigma^{(1)}(x,0)\sigma^{(2)}(x,0)e^{uQ}[\sigma^{(1)^2}(x,0)]\right\}du\Big)dt\Big\} \qquad [5.101]$$

$$-\frac{1}{2K_{10}^{1/2}K_{20}^{3/2}}\Big\{\int_0^T \rho_t\Big(\int_0^t e^{sQ}\left\{e^{tQ}[\sigma^{(1)}(x,0)\sigma^{(2)}(x,0)]\sigma^{(2)^2}(x,0)\right\}ds$$

$$+\int_t^T e^{tQ}\left\{\sigma^{(1)}(x,0)\sigma^{(2)}(x,0)e^{uQ}[\sigma^{(2)^2}(x,0)]\right\}du\Big)dt\Big\},$$

where

$$K_{10} = \int_0^T e^{tQ}(\sigma^{(1)}(x,0))^2dt$$

$$K_{20} = \int_0^T e^{tQ}(\sigma^{(2)}(x,0))^2dt,$$

where Q is the generator of the Markov process $(x_t, \gamma(t))_t$, which is given by

$$Qf(x,t) = \frac{df}{dt}(x,t) + \frac{g_x(t)}{\bar{G}_x(t)}\int_X P(x,dy)[f(y,0) - f(x,t)]. \qquad [5.102]$$

5.6.2. *Appendix 2. Discussions of some extensions*

5.6.2.1. *Current life stochastic volatility driven by a semi-Markov process (current life semi-Markov volatility)*

Let x_t be a semi-Markov process in a measurable phase space $(X, \mathcal{X})$.

We suppose that the stock price S_t satisfies the following stochastic differential equation:

$$dS_t = S_t(rdt + \sigma(x_t, \gamma(t))dw_t)$$

with the volatility $\sigma := \sigma(x_t, \gamma(t))$, depending on the process x_t, which is independent of the standard Wiener process w_t, and the current life $\gamma(t) = t - \tau_{\nu(t)}$, $\mu \in \mathbb{R}$. We call the volatility $\sigma(x_t, \gamma(t))$ the *current life semi-Markov volatility*.

REMARK 5.3.–*We note that process (x_t, γ_t) is a Markov process on $(X, \mathbb{R}_+)$ with infinitesimal operator*

$$Qf(x,t) = \frac{df(x,t)}{dt} + \frac{g_x(t)}{\bar{G}_x(t)} \int_X P(x,dy)[f(y,t) - f(x,t)].$$

5.6.2.2. *Minimal martingale measure*

We consider the following (B, S)-security market (B stands for "Bond" and S stands for "Stock"). Let the stock price S_t satisfy the following equation:

$$dS_t = S_t(\mu dt + \sigma(x_t, \gamma(t))dw_t),$$

where $\mu \in \mathbb{R}$ is the appreciation rate and $\sigma(x_t, \gamma(t))$ is the semi-Markov volatility, and the bond price $B(t)$ is

$$B(t) = B_0 \, e^{rt},$$

where $r > 0$ is the risk-free rate of return (interest rate).

As long as we have two sources of randomness, the Brownian motion $w(t)$ and semi-Markov process x_t, the above (B, S)-security market is incomplete (see theorem 1, Elliott and Swishchuk 2007) and there are many risk-neutral (or martingale) measures. We will also construct the minimal martingale measure (see Föllmer and Schweizer 1991). With respect to this construction (see lemma 4, Elliott and Swishchuk 2007), the following measure P^* is the minimal martingale measure.

Using Girsanov's theorem (see Shiryaev 1999), we obtain the following result concerning the minimal martingale measure in the above market.

LEMMA 5.3.– Under the assumption $\int_0^T \left(\frac{r-\mu}{\sigma(x_t,\gamma(t))}\right)^2 dt < +\infty$, almost surely (a.s.), the following holds:

1) there is a probability measure P^* equivalent to P such that

$$\frac{dP^*}{dP} = \mathbf{E}\left\{\int_0^T \frac{r-\mu}{\sigma(x_t,\gamma(t))} dw(t) - \frac{1}{2}\int_0^T \left(\frac{r-\mu}{\sigma(x_t,\gamma(t))}\right)^2 dt\right\}$$

is its Radon–Nikodym density;

2) the discounted stock price $Z(t) = \frac{S_t}{B(t)}$ is a positive local martingale with respect to P^* and is given by

$$Z(t) = \mathbf{E}\left\{-\frac{1}{2}\int_0^t \sigma^2(x_s,\gamma(s))ds + \int_0^t \sigma(x_s,\gamma(s))dw^*(s)\right\}, \qquad [5.103]$$

where $w^*(t) = \int_0^t \frac{r-\mu}{\sigma(x_s,\gamma(s))}ds + w(t)$ is a standard Brownian motion with respect to P^*.

REMARK 5.4.– *Measure P^* is called the minimal martingale measure.*

REMARK 5.5.– *We note that under risk-neutral measure P^* the stock price S_t satisfies the following equation:*

$$dS_t = S_t(rdt + \sigma(x_t,\gamma(t))dw^*(t))$$

and the discounted process $Z(t)$ has the form

$$dZ(t) = Z(t)\sigma(x_s,\gamma(s))dw^*(t).$$

REMARK 5.6.– *A sufficient condition (Novikov's condition Shiryaev 1999) for the right-hand side of [5.103] to be a martingale is*

$$\mathbf{E}\left\{\frac{1}{2}\int_0^T \left(\frac{r-\mu}{\sigma(x_t,\gamma(t))}\right)^2 dt\right\} < +\infty.$$

5.6.2.3. *Local current stochastic volatility driven by a semi-Markov process (local current semi-Markov volatility)*

We suppose that the stock price S_t satisfies the following stochastic differential equation:

$$dS_t = S_t(rdt + \sigma_{loc}(S_t,x_t,\gamma(t))dw_t)$$

with the volatility $\sigma := \sigma_{loc}(S_t,x_t,\gamma(t))$, depending on the process x_t, which is independent of the standard Wiener process w_t, stock price S_t, and the current life $\gamma(t) = t - \tau_{\nu(t)}$.

REMARK 5.7.– *We note that process (S_t,x_t,γ_t) is a Markov process on $(\mathbb{R}_+,X,\mathbb{R}_+)$ with infinitesimal operator*

$$Qf(s,x,t) = \frac{\partial f(s,x,t)}{\partial t} + \frac{g_x(t)}{\bar{G}_x(t)}\int_X P(x,dy)[f(s,y,t) - f(s,x,t)]$$

$$+ rS\frac{\partial f(s,x,t)}{\partial s} + \frac{1}{2}\sigma^2(s,x,0)S^2\frac{\partial^2 f(s,x,t)}{\partial s^2}. \qquad [5.104]$$

Using the same procedure as in section 5.3.1, with infinitesimal operator Q in [5.104], we can obtain the following result:

THEOREM 5.6.– The value of a variance swap for semi-Markov stochastic volatility $\sigma(x_t)$ is equal to

$$P(x) = e^{-rT}\left(\frac{1}{T}\int_0^T e^{tQ}\sigma^2(s,x,0)dt - K_{var}\right),$$

where Q is defined in [5.104].

5.6.2.4. *Local stochastic volatility driven by a semi-Markov process (local semi-Markov volatility)*

We suppose that the stock price S_t satisfies the following stochastic differential equation:

$$dS_t = S_t(rdt + \sigma(S_t, x_t, t)dw_t)$$

with the volatility $\sigma := \sigma(x_t, t)$ depending on the process x_t, which is independent of the standard Wiener process w_t, stock price S_t and current time t.

Suppose that $\sigma(S, x, t)$ is a differentiable function by t, with a bounded derivative.

Then, we can reduce the problem to the previous one, by the following expansion:

$$\sigma(S_t, x_t, t) = \sigma(S_t, x_t, \gamma(t)) + \tau_{\nu(t)}\frac{d\sigma(S_t, x_t, t)}{dt} + o(\tau_{\nu(t)}).$$

The error of estimation is given by

$$\mathbf{E}[\sigma(S_t, x_t, t) - \sigma(S_t, x_t, \gamma(t))]^2 \leq \mathbf{E}[\tau_{\nu(t)}]^2 \times C,$$

where $C = \max_{0\leq t\leq T}\mathbf{E}\left[\frac{d\sigma(S_t,x_t,t)}{dt}\right]^2$.

5.6.2.5. *Analogue of Dupire formula for semi-Markov local volatility*

Unlike the implied volatility produced by applying the Black–Scholes formula to market prices, the local volatility is the volatility implied by the market prices and the one factor Black–Scholes. Dupire (1994) showed that if the spot price follows a risk-neutral random walk of the form

$$\frac{dS}{S} = (r - D)dt + \sigma(t, S)dW,$$

and if no arbitrage market prices for European vanilla options are available for all strikes K and expires T, then $\sigma(t, S)$ can be extracted analytically from these option prices. If C denotes the price of a European call with strike K and expiry T, then we obtain Dupire's famous equation

$$\frac{\partial C}{\partial T} = \sigma^2_{local}(K, T)\frac{K^2}{2}\frac{\partial^2 C}{\partial K^2} - (r - D)K\frac{\partial C}{\partial K} - DC,$$

where r and D are interest and dividend rates, respectively. After rearranging this equation, we obtain the direct expression to calculate the local volatility (Dupire's formula):

$$\sigma_{local}(K,T) = \sqrt{\frac{\frac{\partial C}{\partial T} + (r-D)K\frac{\partial C}{\partial K} + DC}{\frac{K^2}{2}\frac{\partial^2 C}{\partial K^2}}}. \qquad [5.105]$$

We suppose that the stock price S_t satisfies the following stochastic differential equation:

$$dS_t = S_t((r-D)dt + \sigma_{loc}(S_t, x_t, t)dw_t)$$

where D is the dividend rate, and the volatility $\sigma := \sigma(S_t, x_t, \gamma(t))$ depends on the process x_t, which is independent of the standard Wiener process w_t, stock price and the current life $\gamma(t) = t - \tau_{\nu(t)}$.

Let C denote the price of a European call, with strike K and expiry T, for our (B,S)-security market with semi-Markov volatility. Then, C satisfies the following equation:

$$\frac{\partial C}{\partial T} + (r-D)K\frac{\partial C}{\partial K} - \frac{1}{2}\sigma_{loc}(K,x,T)\frac{\partial^2 C}{\partial K^2} - DC + QC = 0$$

where Q is defined by

$$Qf(x,t) = \frac{g_x(t)}{\bar{G}_x(t)}\int_X P(x,dy)[f(y,t) - f(x,t)]. \qquad [5.106]$$

Then, the semi-Markov local volatility can be calculated using the following expression:

$$\sigma_{loc}(K,x,T) = \sqrt{\frac{\frac{\partial C}{\partial T} + (r-D)K\frac{\partial C}{\partial K} + DC - QC}{\frac{K^2\partial^2 C}{2\partial K^2}}}, \qquad [5.107]$$

where Q is defined in [5.104]. Of course, if σ does not depend on x (and C does not depend as well), then $QC = 0$ in [5.107] coincides with [5.105].

5.6.2.6. *Risk-minimizing strategies (or portfolios) and residual risk*

Let $\pi(t) := (\alpha(t), \beta(t))$ be a *portfolio or strategy* (predictable processes), $V_t(\pi) = \alpha(t) + \beta(t)S_t$ is the *value process*, where S_t satisfies

$$dS_t = S_t\sigma(S_t, x_t, \gamma(t))dw^*(t). \qquad [5.108]$$

We suppose that $r = 0$ for simplicity. Let us define the *cost process*

$$C_t(\pi) = V_t(\pi) - \int_0^t \beta(u)dS_u.$$

The *residual risk* is defined by the following formula (see Föllmer and Schweizer 1991):

$$R_t(\pi) = E[(C_T(\pi) - C_t(\pi))^2 | \mathcal{F}_t].$$

Portfolio $\pi(t)$ is said to be *H-admissible* if $V_T(\pi) = H$. Also, $\pi(t)$ is *self-financing* if the cost process is a martingale. The H-admissible self-financing portfolio π^* is called *risk-minimizing* if, for any H-admissible π and for any t

$$R_t(\pi^*) \leq R_t(\pi).$$

PROPOSITION 5.7 (See Swishchuk 1997).– The risk-minimizing H-admissible strategy $\pi^* = (\alpha^*, \beta^*)$ is given by the following formulas:

$$\beta^*(t) = u_s(S_t, x_t, t)$$

and

$$\alpha^*(t) = V_t(\pi^*) - \alpha^*(t)S_t,$$

where

$$\begin{aligned} V_t(\pi^*) &= E^* f(S_T) + \int_0^t u_s(S_r, x(r), r)dS_r \\ &+ \int_0^t \int_X \phi(r, y)(\mu - \nu)(dr, dy), \end{aligned}$$

$\phi(r, y) := u(S_r, y, r) - u(S_r, x(r^-), r)$, and function $u(z, x, t)$ satisfies the following Cauchy problem:

$$\begin{cases} u_t(z, x, t) + \frac{1}{2}\sigma^2(z, x, t)z^2 u_{zz}(z, x, t) + Qu(z, x, t) = 0 \\ \qquad\qquad\qquad\qquad\qquad\qquad\qquad u(z, xT) = f(z), \end{cases} \quad [5.109]$$

with operator Q as

$$Qu(z, x, t) = \frac{g_x(t)}{\bar{G}_x(t)} \int_X P(x, dy)[u(z, y, t) - u(z, x, t)], \quad [5.110]$$

and where measures μ and ν have been defined in Swishchuk (2010).

PROPOSITION 5.8 (See Swishchuk 1997).– The residual risk process has the form

$$R_t(\pi^*) = \mathbf{E}\left(\int_t^T [Qu^2(S_r, x(r), r) - 2u(S_r, x(r), r)Qu(S_r, x(r), r)]ds | \mathcal{F}_t\right).$$

In particular, the total residual risk on the interval $[0, T]$, at the time $t = 0$, is equal to

$$\begin{aligned} &R_0(\pi^*) \\ &= \mathbf{E}\left(\int_0^T [Qu^2(S_r, x(r), r) - 2u(S_r, x(r), r)Qu(S_r, x(r), r)]ds\right). \end{aligned} \quad [5.111]$$

REMARK 5.8.– *We note that process*

$$m_t := u(S_t, x(t), \gamma(t)) - u(z, x, 0)$$

$$- \int_0^t \left(Q + \frac{d}{dr} \right) u(S_r, x(r), \gamma(r)) dr \qquad [5.112]$$

is an $\mathcal{F}_t$*-martingale, where* $\mathcal{F}_t = \sigma\{x(s), w(s); 0 \leq s \leq t\}$.

PROPOSITION 5.9.– The total residual risk on the interval $[0, T]$ at time $t = 0$ is the risk-neutral expectation of quadratic variation (or characteristic) $\langle m_t \rangle$, of the martingale m_t in [5.105]:

$$R_0(\pi^*) = \mathbf{E}^* \langle m_t \rangle. \qquad [5.113]$$

PROOF.– It follows from the definition of the residual risk, [5.110] and the following relationship:

$$\langle m_t \rangle$$

$$= \int_0^T [(Q + \frac{d}{dr})u^2(S_r, x(r), r) - 2u(S_r, x(r), r)(Q + \frac{d}{dr})u(S_r, x(r), r)]ds$$

$$= \int_0^T [Qu^2(S_r, x(r), r) - 2u(S_r, x(r), r)Qu(S_r, x(r), r)]ds, \qquad [5.114]$$

hence,

$$R_0(\pi^*) = \mathbf{E}^* \langle m_t \rangle,$$

and equation [5.113] follows.

References

Albeverio, S., Korolyuk, V., Samoilenko, I. (2009). Asymptotic expansion of semi-Markov random evolutions. *Stochastics*, 81(5), 477–502.

Angell, J.S. and Olmstead, W.E. (1987). Singularly perturbed Volterra integral equations. *SIAM Journal on Applied Mathematics*, 47(1), 1–14.

Anisimov, V.V. (1977). Switched processes. *Kibernetika*, 4, 111–115.

Arfken, G., Weber, H.J., Harris, F.E. (2012). *Mathematical Methods for Physicists*, 7th edition. Academic Press, Cambridge.

Arratia, R.A. (1979). Coalescing Brownian motions on the line. PhD Thesis, University of Wisconsin, Madison.

Avellaneda, M., Levy, A., Paras, A. (1995). Pricing and hedging derivative securities in markets with uncertain volatility. *Applied Mathematical Finance*, 2, 73–88.

Balakrishnan, V., Van den Broeck, C., Hänggi, P. (1988). First-passage of non-Markovian processes: The case of a reflecting boundary. *Physical Review A*, 38(8), 4213–4222.

Banchoff, T. (1976). Self linking numbers of space polygons. *Indiana University Mathematics Journal*, 25(12), 1171–1188.

Bateman, H. (1955). *Higher Transcendental Functions*. McGraw-Hill, New York.

Bateman, H. (1954). *Tables of Integral Transforms*. McGraw-Hill, New York.

Beghin, L. and Orsingher, E. (2010a). Moving randomly amid scattered obstacles. *Stochastics*, 82, 201–229.

Beghin, L. and Orsingher, E. (2010b). Poisson-type processes governed by fractional and higher-order recursive differential equations. *Electronic Journal of Probability*, 15, 684–709 [Online]. Available at: https://doi.org/10.1214/EJP.v15-762.

Benth, F.E., Groth, M., Kufakunesu, R. (2007). Valuing volatility and variance swaps for a non-Gaussian Ornstein–Uhlenbeck stochastic volatility model. *Applied Mathematical Finance*, 14(4), 347–363.

Bertoin, J. (2002). Self-attracting Poisson clouds in an expanding universe. *Communications in Mathematical Physics*, 232, 59–81.

Black, F. and Scholes, M. (1973). The pricing of options and corporate liabilities. *Journal of Political Economy*, 81, 637–54.

Bochner, S. and Chandrasekhar, K. (1949). *Fourier Transforms, Volume 19: Annals of Mathematics Studies*. Princeton University Press, Princeton.

Bollerslev, T. (1986). Generalized autoregressive conditional heteroscedasticity. *Journal of Economics*, 31, 307–327.

Bossu, S. (2005). Arbitrage pricing of equity correlation swaps. Working paper, J.P. Morgan Equity Derivatives, London.

Bossu, S. (2007). A new approach for modelling and pricing correlation swaps. Equity Structuring. Working paper, ECD, London.

Broadie, M. and Jain, A. (2008a). The effect of jumps and discrete sampling on volatility and variance swaps. *International Journal of Theoretical and Applied Finance*, 11(8), 761–797.

Broadie, M. and Jain, A. (2008b). Pricing and hedging volatility derivatives. *The Journal of Derivatives*, 15(3), 7–24.

Brockhaus, O. and Long, D. (2000). Volatility swap made simple. *Risk*, 2(1), 92–96.

Buff, R. (2002). *Uncertain Volatility Model. Theory and Applications*. Springer, New York.

Cahoy, D.O. (2007). Fractional Poisson process in terms of α-stable densities. PhD Thesis, Case Western University, Cleveland.

Carr, P. and Lee, R. (2007). Realized volatility and variance: Options via swaps. Technical report, Bloomberg LP and University of Chicago [Online]. Available at: http://math.uchicago.edu/~rl/OVSwithAppendices.pdf.

Carr, P. and Lee, R. (2009). Volatility derivatives. *Annual Review of Financial Economics*, 1, 319–39.

Carr, P., Geman, H., Madan, D.B., Yor, M. (2005). Pricing options on realized variance. *Finance and Stochastics*, 9, 453–475.

Chesney, M. and Scott, L. (1989). Pricing European currency options: A comparison of modified Black–Scholes model and a random variance model. *Journal of Financial and Quantitative Analysis*, 24(3), 267–284.

Cox, D.R. (1970). *Renewal Theory*. Methuen & Co., London.

Cox, J. and Ross, R. (1976). Valuation of options for alternative stochastic processes. *Journal of Financial Economics*, 3(1), 145–166.

Cox, J., Ross, S., Rubinstein, M. (1979). Option pricing: A simplified approach. *Journal of Financial Economics*, 7, 229–263.

Da Fonseca, J., Ielpo, F., Grasselli, M. (2009). Hedging (co)variance risk with variance swaps [Online]. Available at: http://ssrn.com/abstract=1341811.

De Gregorio, A. (2012). On random flights with non-uniformly distributed directions. *Journal of Statistical Physics*, 147(2), 382–411.

De Gregorio, A. (2014). A family of random walks with generalized Dirichlet steps. *Journal of Mathematical Physics*, 55, 023302.

De Gregorio, A. and Orsingher, E. (2012). Flying randomly in $\mathbb{R}^d$ with Dirichlet displacements. *Stochastic Processes and their Applications*, 122, 676–713.

De Gregorio, A., Orsingher, E., Sakhno, L. (2005). Motions with finite velocity analyzed with order statistics and differential equations. *Theory of Probability and Mathematical Statistics*, 71, 63–79.

Demeterfi, K., Derman, E., Kamal, M., Zou, J. (1999). A guide to volatility and variance swaps. *The Journal of Derivatives*, 6(4), 9–32.

Derman, E. (1999). Regimes of volatility. *Risk*, 4, 55–59.

Derman, E. and Kani, I. (1994). Riding on a smile. *Risk*, 7(2), 32–39.

Di Crescenzo, A. (2001). On random motions with velocities alternating at Erlang-distributed random times. *Advances in Applied Probability*, 33, 690–701.

Di Masi, G.B., Kabanov, Y.M., Runggaldier, W.J. (1994). Hedging of options on stock under mean-square criterion and Markov volatilities. *Theory Probability and Its Applications*, 39(1), 211–222.

Dorogovtsev, A.A. (2007a). Fourier-Wiener transforms of functionals on arratia flow. *Ukrainian Mathematical Bulletin*, 4(3), 333–354.

Dorogovtsev, A.A. (2007b). Measure-valued stochastic processes and flows. Institute of Mathematics NAS of Ukraine, Kyiv.

Dorogovtsev, A.A. (2010). One Brownian stochastic flow. *Theory of Stochastic Processes*, 10, 21–25.

D'Ovidio, M., Orsingher, E., Toaldo, B. (2014). Fractional telegraph-type equations and hyperbolic Brownian motion. *Statistics & Probability Letters*, 89, 131–137.

Drissien, J., Maenhout, P.J., Vilkov, G. (2009). The price of correlation risk: Evidence from equity options. *The Journal of Finance*, 64(3), 1377–1406.

Duan, J. (1995). The Garch option pricing model. *Mathematical Finance*, 5, 13–32.

Dupire, B. (1994). Pricing with a smile. *Risk*, 7(1), 18–20.

Economou, A. (2002). An alternative model for queueing systems with single arrivals, batch services and customer coalescence. *Queueing Systems*, 40, 407–432.

Elliott, R. and Swishchuk, A.V. (2007). Pricing options and variance swaps in Markov-modulated Brownian markets. In *Hidden Markov Models in Finance*, Mamon, R., Elliott, R. (eds), Springer, New York.

Elliott, R.J., Siu, T.K., Chan, L. (2007a). Pricing options under a generalized Markov-modulated jump-diffusion model. *Stochastic Analysis and Applications*, 25, 821–843.

Elliott, R.J., Siu, T.K., Chan, L. (2007b). Pricing volatility swaps under Heston's stochastic volatility model with regime switching. *Applied Mathematical Finance*, 14(1), 41–62.

Feller, W. (1968). *An Introduction to Probability Theory and Its Applications*, Volume 1, 3rd edition. John Wiley & Sons, New York.

Feller, W. (1971). *An Introduction to Probability Theory and Its Applications*, Volume 2, 2nd edition. John Wiley & Sons, New York.

Föllmer, H. and Schweizer, M. (1991). Hedging of contingent claims under incomplete information. In *Applied Stochastic Analysis Volume 5*, Davis, M.H.A., Elliott, R.J. (eds). Gordon and Breach, London.

Föllmer, H. and Sondermann, D. (1986). Hedging of non-redundant contingent claim. In *Contributions to Mathematical Economics*, Hilenbrandt, W., Mas-Colell, A. (eds). North-Holland, Amsterdam.

Franceschetti, M. (2007). When a random walk of fixed length can lead uniformly anywhere inside a hypersphere. *Journal of Theoretical Probability*, 20, 813–823.

Frey, R. (1997). Derivative asset analysis in models with level-dependent and stochastic volatility. *CWI Quarterly*, 10, 1–34.

Gakhov, F.D. and Cherskii, Y.I. (1978). *Convolution Type Equations* [in Russian]. Nauka. Moscow.

Garra, R. and Orsingher, E. (2014). Random flights governed by Klein-Gordon-type partial differential equations. *Stochastic Processes and their Applications*, 124, 2171–2187.

Giraud, C. (2001). Clustering in a self-gravitating one-dimensional gas at zero temperature. *Journal of Statistical Physics*, 105, 585–604.

Giraud, C. (2005). Gravitational clustering and additive coalescence. *Stochastic Processes and their Applications*, 115, 1302–1322.

Girko, V.L. (1982). Limit theorems for products of random matrices with positive elements [in Russian]. *Teor. Veroyatnost. i Primenen.*, 27, 777–783.

Goldstein, S. (1951). On diffusion by discontinuous movements and on the telegraph equation. *The Quarterly Journal of Mechanics and Applied Mathematics*, 4, 129–156.

Gorostiza, L.G. (1973). The central limit theorem for random motions of d-dimensional Euclidean space. *The Annals of Probability*, 1(4), 603–612.

Gorostiza, L.G. and Griego, R.J. (1979). Strong approximations of diffusion processes by transport processes. *Kyoto Journal of Mathematics*, 19(1), 91–103.

Gradshteyn, I.S. and Ryzhik, I.M. (1980). *Tables of Integrals, Sums, Series and Products*. Academic Press, San Diego.

Gray, S. (1996). Modeling the conditional distribution of interest rates as a regime-switching process. *Journal of Financial Econometrics*, 42, 27–62.

Griego, R. and Hersh, R. (1969). Random evolutions, Markov chains, and systems of partial differential equations. *Proceedings of the National Academy of Sciences*, 62, 305–308.

Griego, R. and Hersh, R. (1971). Theory of random evolutions and applications to partial differential equations. *Transactions of the American Mathematical Society*, 156, 405–418.

Griego, R. and Korzeniowski, A. (1989). On principal eigenvalues for random evolutions. *Stochastic Analysis and Applications*, 7, 35–45.

Griego, R. and Swishchuk, A.V. (2000). Black-Scholes formula for a market in a Markov environment. *Theory of Probability and Mathematical Statistics*, 62, 9–18.

Hamilton, J. (1989). Rational-expectations econometric analysis of changes in regime. *Journal of Economic Dynamics and Control*, 12, 385–423.

Harrison, J. and Pliska, S. (1981). Martingales, stochastic integrals in the theory of continuous trading. *Stochastic Processes and their Applications*, 11(3), 215–260.

Hersh, R. (1974). Random evolutions: A survey of results and problems. *Rocky Mountain Journal of Mathematics*, 4, 443–477.

Hersh, R. (2003). The birth of random evolutions. *The Mathematical Intelligencer*, 25(1), 53–60.

Hersh, R. and Papanicolaou, G. (1972). Non-commuting random evolutions, and an operator-valued Feynman–Kac formula. *Communications on Pure and Applied Mathematics*, 30, 337–367.

Hersh, R. and Pinsky, M. (1972). Random evolutions are asymptotically Gaussian. *Communications on Pure and Applied Mathematics*, 25, 33–44.

Heston, S. (1993). A closed-form solution for options with stochastic volatility with applications to bond and currency options. *Review of Financial Studies*, 6, 327–343.

Hobson, D. and Rogers, L. (1998). Complete models with stochastic volatility. *Mathematical Finance*, 8(1), 27–48.

Howison, S., Rafailidis, A., Rasmussen, H. (2004). On the pricing and hedging of volatility derivatives. *Applied Mathematical Finance*, 11, 317–346.

Hull, J. (2000). *Options, Futures and Other Derivatives*, 4th edition. Prentice Hall, Upper Saddle River.

Hull, J. and White, A. (1987). The pricing of options on assets with stochastic volatilities. *The Journal of Finance*, 42, 281–300.

Ignatovich, S.R., Kucher, A.G., Yakushenko, A.S., Bashta, A.V. (2004). Modeling of coalescence of dispersed surface cracks. *Strength of Materials*, 36(2), 125–133.

Iksanov, A.M. and Rösler, U. (2006). Some moment results about the limit of a martingale related to the supercritical branching random walk and perpetuities. *Ukrainian Mathematical Journal*, 58(4), 451–471.

Javaheri, A., Wilmott, P., Haug, E. (2002). Garch and volatility swaps. Technical article, WILMOTT.

Johnson, H. and Shanno, D. (1987). Option pricing when the variance is changing. *Journal of Financial and Quantitative Analysis*, 22, 143–151.

Kac, M. (1974). A stochastic model related to the telegrapher's equation. *Rocky Mountain Journal of Mathematics*, 4, 497–509.

Kazmerchuk, Y., Swishchuk, A.V., Wu, J. (2002). A continuous-time Garch model for stochastic volatility with delay. *Canadian Applied Mathematics Quarterly*, 13, 2.

Kelvin, B. and Thomson, W. (1854). On the theory of the electric telegraph. *Proceedings of the Royal Society of London*, 7, 382–399.

Kertz, R. (1978). Random evolutions with underlying semi-Markov processes. *Publications of the Research Institute for Mathematical Sciences*, 14(3), 589–614.

Kolesnik, A.D. and Pinsky, M.A. (2011). Random evolutions are driven by the hyperparabolic operators. *Journal of Statistical Physics*, 142, 828–846.

Kolesnik, A. and Ratanov, N. (2013). *Telegraph Processes and Option Pricing*. Springer, Heidelberg.

Konarovskii, V.V. (2011). On infinite system of diffusing particles with coalescing. *Theory of Probability and Its Applications*, 55(1), 134–144.

Korn, G.A. and Korn, T.M. (2000). *Mathematical Handbook for Scientists and Engineers*, 2nd edition. Dover Publications, New York.

Kornfeld, I.P., Sinai, Y.G., Fomin, S.V. (1982). *Ergodic Theory*. Springer-Verlag, Berlin.

Korolyuk, V.S. and Korolyuk, V.V. (1999). *Stochastic Models of Systems*. Kluwer Academic Publishers, Dordrecht.

Korolyuk, V.S. and Limnios, N. (2004). Average and diffusion approximation for evolutionary systems in an asymptotic split phase space. *The Annals of Applied Probability*, 14(1), 489–516.

Korolyuk, V.S. and Limnios, N. (2005). *Stochastic Systems in Merging Phase Space*. World Scientific Publishing Company, New Jersey.

Korolyuk, V.S. and Limnios, N. (2009). Poisson approximation of processes with locally independent increments with Markov switching. *Theory of Stochastic Processes*, 15(1), 40–48.

Korolyuk, V.S. and Swishchuk, A.V. (1986). The central limit theorem for semi-Markov random evolutions. *Ukrainian Mathematical Journal*, 38, 286–289.

Korolyuk, V.S. and Swishchuk, A.V. (1995). *Semi-Markov Random Evolution*. Kluwer Academic Publishers, Dordrecht.

Korolyuk, V.S. and Turbin, A.F. (1976). *Semi-Markov Processes and their Applications*. Naukova Dumka, Kyiv.

Korolyuk, V.S. and Turbin, A.F. (1993). *Mathematical Foundations of the State Lumping of Large Systems*. Kluwer Academic Publishers, Dordrecht.

Kovalenko, I.N., Kuznetsov, N.Y., Shurenkov, V.M. (1983). *Random Processes: Manual*. Naukova Dumka, Kyiv.

Krein, M.G. (1962). Integral equations on a half-line. In *Nine papers on analysis*, Ahiezer, N.I., Ašnevic, I.A., Efimov, N.V., Kreĭn, M.G., Lebedev, N.A., Markuševič, A.I., Sidlovskiĭ, A.B., Šmul'jan, J.L., Ulina, G.V. (eds). American Mathematical Society Translations: Series 2, Providence, Rhode Island.

Lachal, A. (2006). Cyclic random motions in $\mathbb{R}^d$-space with n directions. *ESAIM: Probability and Statistics*, 10, 277–316.

Le, J.Y. and Raimond, O. (2004). Flows, coalescence and noise. *The Annals of Probability*, 32(2), 1247–1315.

Le Caër, G. (2010). A Pearson-Dirichlet random walk. *Journal of Statistical Physics*, 140, 728–751.

Le Caër, G. (2011). A new family of solvable Pearson-Dirichlet random walks. *Journal of Statistical Physics*, 144, 23–45.

Letac, G. and Piccioni, M. (2014). Dirichlet random walks. *Journal of Applied Probability*, 51, 1081–1099.

Lifshits, M. and Shi, Z. (2005). Aggregation rates in one-dimensional stochastic systems with adhesion and gravitation. *The Annals of Probability*, 33(1), 53–81.

López, O. and Ratanov, N. (2012). Option pricing under jump-telegraph model with random jumps. *Journal of Applied Probability*, 49(3), 838–849.

Marcus, A.H. (1968). Stochastic coalescence. *Technometrics*, 10(1), 133–143.

Masoliver, J., Porra, J.M., Weiss, G.H. (1993a). Solution to the telegrapher's equation in the presence of reflecting and partly reflecting boundaries. *Physical Review*, 48(2), 939–944.

Masoliver, J., Porrá, J.M., Weiss, G.H. (1993b). Some two and three-dimensional persistent random walks. *Physica A*, 193, 469–482.

Merton, R. (1973). Theory of rational option pricing. *Bell Journal of Economic Management Science*, 4, 141–183.

Meyer, P. (1969). Processus à accroissements indépendants et positifs. *Séminaire de probabilitéss (Strasbourg)*, 3, 175–189.

Morse, P.M. and Feshbach, H. (1953). *Methods of Theoretical Physics*. McGraw-Hill, New York.

Nischenko, I.I. (2001). On asymptotic representation normalizing multiplier for random matrix-valued evolution. *Theory of Probability and Mathematical Statistics*, 64, 129–135.

Orsingher, E. (1985). Hyperbolic equations arising in random models. *Stochastic Processes and their Applications*, 21, 93–106.

Orsingher, E. and Beghin, L. (2006). *Probabiliti e modelli aleatori*. Aracne, Rome.

Orsingher, E. and Beghin, L. (2009). Fractional diffusion equations and processes with randomly-varying time. *Annals of Probability*, 37(1), 206–249.

Orsingher, E. and De Gregorio, A. (2007). Random flights in higher spaces. *Journal of Theoretical Probability*, 20, 769–806.

Orsingher, E. and Ratanov, N. (2002). Planar random motions with drift. *Journal of Applied Mathematics and Stochastic Analysis*, 15, 205–221.

Orsingher, E. and Ratanov, N. (2007). Exact distributions of random motions in inhomogeneous media. *Theory of Probability and Mathematical Statistics*, 76, 125–137.

Orsingher, E. and Somella, A.M. (2004). A cyclic random motion in $\mathbb{R}^3$ with four directions and finite velocity. *Stochastics and Stochastics Reports*, 76, 113–133.

Papanicolaou, G. (1971a). Asymptotic analysis of transport processes. *Bulletin of the American Mathematical Society*, 81, 330–391.

Papanicolaou, G. (1971b). Motion of a particle in a random field. *Journal of Mathematical Physics*, 12(24), 473–489.

Pilipenko, A.Y. (1999). The evolution of a system of particles and measure-valued processes. *Theory of Stochastic Processes*, 5(3–4), 188–197.

Pinsky, M. (1991). *Lectures on Random Evolutions*. World Scientific, Singapore.

Pogorui, A.A. (1989). Distribution of reaching time of infinitely increasing level by family of semi-Markov processes with extending phase space [in Russian]. In *Analytical Methods of Research of Evolution Stochastic Systems*, Institute of Mathematics of NAS of Ukraine, Kyiv.

Pogorui, A.A. (1994). Asymptotic inequalities for the distribution of the time of stay of a semi-Markov process in an expanding set of states. *Ukrainian Mathematical Journal*, 46(11), 1757–1762.

Pogorui, A.A. (2003). Estimation of the stationary efficiency of a two-phase system with two storages. *Journal of Automation and Information Sciences*, 35, 16–23.

Pogorui, A.A. (2004). Estimation of the efficiency of a data transmission line with feedback. *Journal of Automation and Information Sciences*, 36, 44–50.

Pogorui, A.A. (2005). Stationary distribution of the random shift process with delaying in reflecting boundaries. *Scientific Journal of National Pedagogical Dragomanov University, Series 1*, 6, 168–172.

Pogorui, A.A. (2006). Stationary distribution of a process of random semi-Markov evolution with delaying screens in the case of balance. *Ukrainian Mathematical Journal*, 58(3), 430–437.

Pogorui, A.A. (2007). Hyperholomorphic functions in commutative algebras. *Complex Variables and Elliptic Equations*, 52(12), 1155–1159.

Pogorui, A.A. (2009a). Asymptotic expansion for distribution of Markovian random motion. *Random Operators and Stochastic Equations*, 17(2), 189–196.

Pogorui, A.A. (2009b). Stationary distributions of fading evolutions. *Ukrainian Mathematical Journal*, 61(3), 425–431.

Pogorui, A.A. (2010a). Asymptotic analysis for phase averaging of transport process. *Ukrainian Mathematical Journal*, 62(2), 190–198.

Pogorui, A.A. (2010b). Fading evolution in multidimensional spaces. *Ukrainian Mathematical Journal*, 62(11), 1577–1582.

Pogorui, A.A. (2011). The distribution of random evolution in Erlang semi-Markov media. *Theory of Stochastic Processes*, 17(33), 90–99.

Pogorui, A.A. (2012a). Evolution in multidimensional spaces. *Random Operators and Stochastic Equations*, 20(2), 135–141.

Pogorui, A.A. (2012b). System of interactive particles with Markovian switching. *Theory of Stochastic Processes*, 18(34–2), 83–95.

Pogorui, A.A. and Rodríguez-Dagnino, R.M. (2005). One-dimensional semi-Markov evolution with general Erlang sojourn times. *Random Operators and Stochastic Equations*, 13(4), 399–405.

Pogorui, A.A. and Rodríguez-Dagnino, R.M. (2006). Limiting distribution of random motion in an *n*-dimensional parallelepiped. *Random Operators and Stochastic Equations*, 14(4), 385–392.

Pogorui, A.A. and Rodríguez-Dagnino, R.M. (2009a). Evolution process as an alternative to diffusion process and Black–Scholes formula. *Random Operators and Stochastic Equations*, 17(1), 61–68.

Pogorui, A.A. and Rodríguez-Dagnino, R.M. (2009b). Limiting distribution of fading evolution in some semi-Markov media. *Ukrainian Mathematical Journal*, 61(12), 1720–1724.

Pogorui, A.A. and Rodríguez-Dagnino, R.M. (2010a). Asymptotic expansion for transport processes in semi-Markov media. *Theory of Probability and Mathematical Statistics*, 83, 127–134.

Pogorui, A.A. and Rodríguez-Dagnino, R.M. (2010b). Stationary distribution of random motion with delay in reflecting boundaries. *Applied Mathematics*, 1(1), 24–28.

Pogorui, A.A. and Rodríguez-Dagnino, R.M. (2011). Isotropic random motion at finite speed with k-Erlang distributed direction alternations. *Journal of Statistical Physics*, 145(1), 102–112.

Pogorui, A.A. and Rodríguez-Dagnino, R.M. (2012). Random motion with uniformly distributed directions and random velocity. *Journal of Statistical Physics*, 147(6), 1216–1225.

Pogorui, A.A. and Rodríguez-Dagnino, R.M. (2013). Random motion with gamma steps in higher dimensions. *Statistics & Probability Letters*, 83, 1638–1643.

Pogorui, A.A. and Turbin, A.F. (2002). Estimation of stationary efficiency of a production line with two unreliable aggregates. *Cybernetics and Systems Analysis*, 38, 823–829.

Pogorui, A.A., Rodríguez-Dagnino, R.M., Shapiro, M. (2014). Solutions for PDEs with constant coefficients and derivability of functions ranged in commutative algebras. *Mathematical Methods in the Applied Sciences*, 37(17), 2799–2810.

Prudnikov, A.P., Brychkov, Y., Marichev, O.I. (1986). *Integrals and Series: More Special Functions*. Gordon and Breach, London.

Qian, H., Raymond, G.M., Bassingthwaighte, J.B. (1998). On two-dimensional fractional Brownian motion and fractional Brownian random field. *Journal of Physics A: Mathematical and General*, 31, 1527–1535.

Ratanov, N. (2007). A jump telegraph model for option pricing. *Quantitative Finance*, 7(5), 575–583.

Ratanov, N. (2010). Option pricing model based on a Markov-modulated diffusion with jumps. *The Brazilian Journal of Probability and Statistics*, 24, 413–431.

Ratanov, N. and Melnikov, A. (2008). On financial markets based on telegraph processes. *Stochastics*, 80, 247–268.

Rodríguez-Said, R.D., Pogorui, A.A., Rodríguez-Dagnino, R.M. (2007). Stationary probability distribution of a system with n equal customers with bursty demands connected to a single buffer. *Random Operators and Stochastic Equations*, 15(2), 181–204.

Salvi, G. and Swishchuk, A.V. (2012a). Covariance and correlation swaps for financial markets with Markov-modulated volatility model. *International Journal of Theoretical and Applied Finance*, 17, 1.

Salvi, G., Swishchuk, A.V. (2012b). *Pricing of Variance, Volatility, Covariance and Correlation Swaps in a Markov-modulated Volatility Model*. University of Calgary, Calgary.

Samoilenko, I.V. (2001). Markovian random evolution in $\mathbb{R}^n$. *Random Operators and Stochastic Equations*, 9, 247–257.

Samoilenko, I.V. (2002). Fading Markov random evolution. *Ukrainian Mathematical Journal*, 54(3), 448–459.

Samoilenko, I.V. (2005). Asymptotic expansion for the functional of Markovian evolution in $\mathbb{R}^d$ in the circuit of diffusion approximation. *Journal of Applied Mathematics and Stochastic Analysis*, 3, 247–257.

Scott, L. (1987). Option pricing when the variance changes randomly: Theory, estimation and an application. *Journal of Financial and Quantitative Analysis*, 22, 419–438.

Sepp, A. (2008). Pricing options on realized variance in the Heston model with jumps in returns and volatility. *Journal of Computational Finance*, 11(4), 33–70.

Shiryaev, A. (1999). *Essentials of Stochastic Finance: Facts, Models, Theory*. World Scientific, Singapore.

Shreve, S. (2004). *Stochastic Calculus for Finance: Continuous-time Models*. Springer, New York.

Shubin, C. (2006). Singularly perturbed integral equations. *Journal of Mathematical Analysis and Applications*, 313(1), 234–250.

Shurenkov, V.M. (1986). Markov intervention of chance and limit theorems. *Mathematics of the USSR-Sbornik*, 54, 1.

Shurenkov, V.M. (1989). *Ergodic Markov Processes* [in Russian]. Nauka, Moscow.

Sinai, Y.G. (1992). Distribution of some functionals of the integral of a random walk. *Theoretical and Mathematical Physics*, 90, 219–241.

Sinai, Y.G. (2010). *Dynamical Systems, Ergodic Theory and Applications, Volume 100*. Encyclopedia of Mathematical Sciences, Springer-Verlag, Berlin.

Skorokhod, A.V. (1989). *Asymptotic Methods in the Theory of Stochastic Differential Equations*. American Mathematical Society, Rhode Island.

Stadje, W. (2007). The exact probability distribution of a two-dimensional random walk. *Journal of Statistical Physics*, 56, 207–216.

Stadje, W. and Zacks, S. (2004). Telegraph processes with random velocities. *Journal of Applied Probability*, 41, 665–678.

Stein, E. and Stein, J. (1991). Stock price distributions with stochastic volatility: An analytic approach. *The Review of Financial Studies*, 4, 727–752.

Stroock, D.W. and Varadhan, S.R.S. (1969). Diffusion processes with continuous coefficients, *Communications on Pure and Applied Mathematics*, 22(3), 345–400.

Stroock, D.W. and Varadhan, S.R.S. (1979). *Multidimensional Diffusion Processes*. Springer-Verlag, Berlin.

Sviridenko, M.N. (1986). Martingale approach to limit theorems for semi-Markov processes. *Theory of Probability and Its Applications*, 540–545.

Swishchuk, A.V. (1989). Weak convergence of semi-Markov random evolutions in an averaging scheme (martingale approach). *Ukrainian Mathematical Journal*, 41(3), 1450–1456.

Swishchuk, A.V. (1997). Hedging of options under mean-square criterion and with semi-Markov volatility. *Ukrainian Mathematical Journal*, 47(7), 1119–1127.

Swishchuk, A.V. (2000). *Random Evolutions and their Applications: New Trends*. Kluwer Academic Publishers, Dordrecht.

Swishchuk, A.V. (2004). Modeling of variance and volatility swaps for financial markets with stochastic volatilities. *WILMOTT Magazine*, 2, 64–72.

Swishchuk, A.V. (2005). Modeling and pricing of variance swaps for stochastic volatilities with delay. *WILMOTT Magazine*, 19, 63–73.

Swishchuk, A.V. (2010). Pricing of variance and volatility swaps with semi-Markov volatilities. *Canadian Applied Mathematics Quarterly*, 18, 4.

Swishchuk, A.V., Cheng, R., Lawi, S., Badescu, A., Mekki, H.B., Gashaw, A.F., Hua, Y., Molyboga, M., Neocleous, T., Petrachenko, Y. (2002). Price pseudo-variance, pseudo-covariance, pseudo-volatility, and pseudo-correlation swaps – In analytical closed-forms. In *Proceedings of the Sixth PIMS Industrial Problems Solving Workshop*, Macki, J. (ed.). University of Alberta, Edmonton.

Tautz, R.C. and Lerche, I. (2016). Application of the three-dimensional telegraph equation to cosmic-ray transport. *Research in Astronomy and Astrophysics*, 16, 10.

Théoret, R., Zabré, L., Rostan, P. (2002). Pricing volatility swaps: Empirical testing with Canadian data. Working paper 17, Centre de Recherche en Gestion, Montreal.

Tóth, B. and Werener, W. (1998). The true self-repelling motion. *Probability Theory and Related Fields*, 111, 375–452.

Turbin, A.F. (1972). The solution of some problems on perturbing the occurrence of matrices. *Reports of the National Academy of Sciences of Ukraine Series A*, 539–541.

Turbin, A.F. (1981). Limit theorems for perturbed semigroups and Markov processes in the scheme of asymptotic phase lumping [in Russian]. Institute of Mathematics, Ukrainian National Academy of Sciences, Preprint no. 80.18. 133–147.

Turbin, A.F. (1998). Mathematical models of one-dimensional Brownian motion as alternatives to the mathematical models of Einstein, Wiener, Levy [in Russian]. *Fractal Analysis and Related Fields*, 2, 47–60.

Vladimirov, V.S. (1996). *Equations of Mathematical Physics*. Mir Publishers, Moscow.

Vysotsky, V.V. (2008a). Clustering in a stochastic model of one-dimensional gas. *The Annals of Applied Probability*, 18(3), 1026–1058.

Vysotsky, V.V. (2008b). Limit theorems for stochastic models of interacting particles. PhD Thesis, Saint Petersburg State University, Saint Petersburg.

Watkins, J. (1984). A central limit theorem in random evolution. *The Annals of Probability*, 12(2), 480–514.

Watkins, J. (1985). Limit theorems for stationary random evolutions. *Stochastic Processes and their Applications*, 19, 189–224.

Wiggins, J. (1987). Option values under stochastic volatility: Theory and empirical estimates. *Journal of Financial Economics*, 19, 351–372.

Wilmott, P., Howison, S., Dewynne, J. (1995). *Option Pricing: Mathematical Models and Computations*. Oxford Financial Press, Oxford.

Windcliff, H., Forsyth, P.A., Vetzal, K.R. (2006). Pricing methods and hedging strategies for volatility derivatives. *Journal of Banking & Finance*, 30, 409–431.

Yeleyko, Y.I. and Zhernovyi, Y.V. (2002). Asymptotic properties of stochastic evolutions described by the solutions of the ordinary differential equations partially solved with respect to the higher derivatives. *Bulletin of the Lviv University, Series: Physics and Mathematics*, 60, 60–66.

Index

B, D, G

Black–Scholes formula, 101, 106–108, 112–114, 172
correlation swaps, 111, 116, 118, 128, 130, 132, 135, 138, 143, 144, 147, 159, 160, 162, 164
covariance swaps, 128, 143, 146, 159, 160
Dupire formula, 112, 172, 173
Goldstein–Kac
- equation in three dimensions, 40
- process, 21

M, S

Markov-modulated
- stochastic volatility, 118, 125
- volatility, 111

motion with random velocity, 4, 17
semi-Markov
- media, 17, 32
- modulated volatility, 153, 158
- volatility, 143, 144, 147, 169, 170, 173

sphere of singularity, 16, 17

T, V

telegraph equation, 21
three-dimensional random motion, 13, 32
variance swaps, 115–118, 124, 128, 145–147, 153
volatility, 106, 111–115, 124, 143
- skew, 112, 115
- smile, 112
- swaps, 111, 115–118, 122, 124, 125, 134

Volterra renewal-type integral equation, 5, 15, 19

Summary of Volume 1

Preface

Acknowledgments

Introduction

Part 1. Basic Methods

Chapter 1. Preliminary Concepts

1.1. Introduction to random evolutions
1.2. Abstract potential operators
1.3. Markov processes: operator semigroups
1.4. Semi-Markov processes
1.5. Lumped Markov chains
1.6. Switched processes in Markov and semi-Markov media

Chapter 2. Homogeneous Random Evolutions (HRE) and their Applications

2.1. Homogeneous random evolutions (HRE)
2.1.1. Definition and classification of HRE
2.1.2. Some examples of HRE
2.1.3. Martingale characterization of HRE
2.1.4. Analogue of Dynkin's formula for HRE
2.1.5. Boundary value problems for HRE
2.2. Limit theorems for HRE
2.2.1. Weak convergence of HRE
2.2.2. Averaging of HRE

2.2.3. Diffusion approximation of HRE
2.2.4. Averaging of REs in reducible phase space: merged HRE
2.2.5. Diffusion approximation of HRE in reducible phase space
2.2.6. Normal deviations of HRE
2.2.7. Rates of convergence in the limit theorems for HRE

Part 2. Applications to Reliability, Random Motions, and Telegraph Processes

Chapter 3. Asymptotic Analysis for Distributions of Markov, Semi-Markov and Random Evolutions

3.1. Asymptotic distribution of time to reach a level that is infinitely increasing by a family of semi-Markov processes on the set $\mathbb{N}$
3.2. Asymptotic inequalities for the distribution of the occupation time of a semi-Markov process in an increasing set of states
3.3. Asymptotic analysis of the occupation time distribution of an embedded semi-Markov process (with increasing states) in a diffusion process
3.4. Asymptotic analysis of a semigroup of operators of the singularly perturbed random evolution in semi-Markov media
3.5. Asymptotic expansion for distribution of random motion in Markov media under the Kac condition
3.5.1. The equation for the probability density of the particle position performing a random walk in $\mathbb{R}^n$
3.5.2. Equation for the probability density of the particle position
3.5.3. Reduction of a singularly perturbed evolution equation to a regularly perturbed equation
3.6. Asymptotic estimation for application of the telegraph process as an alternative to the diffusion process in the Black–Scholes formula
3.6.1. Asymptotic expansion for the singularly perturbed random evolution in Markov media in case of disbalance
3.6.2. Application to an economic model of stock market

Chapter 4. Random Switched Processes with Delay in Reflecting Boundaries

4.1. Stationary distribution of evolutionary switched processes in a Markov environment with delay in reflecting boundaries
4.2. Stationary distribution of switched process in semi-Markov media with delay in reflecting barriers

4.2.1. Infinitesimal operator of random evolution with semi-Markov switching
4.2.2. Stationary distribution of random evolution in semi-Markov media with delaying boundaries in balance case
4.2.3. Stationary distribution of random evolution in semi-Markov media with delaying boundaries
4.3. Stationary efficiency of a system with two unreliable subsystems in cascade and one buffer: the Markov case
4.3.1. Introduction
4.3.2. Stationary distribution of Markov stochastic evolutions
4.3.3. Stationary efficiency of a system with two unreliable subsystems in cascade and one buffer
4.3.4. Mathematical model
4.3.5. Main mathematical results
4.3.6. Numerical results for the symmetric case
4.4. Application of random evolutions with delaying barriers to modeling control of supply systems with feedback: the semi-Markov switching process
4.4.1. Estimation of stationary efficiency of one-phase system with a reservoir
4.4.2. Estimation of stationary efficiency of a production system with two unreliable supply lines

Chapter 5. One-dimensional Random Motions in Markov and Semi-Markov Media

5.1. One-dimensional semi-Markov evolutions with general Erlang sojourn times
5.1.1. Mathematical model
5.1.2. Solution of PDEs with constant coefficients and derivability of functions ranged in commutative algebras
5.1.3. Infinite-dimensional case
5.1.4. The distribution of one-dimensional random evolutions in Erlang media
5.2. Distribution of limiting position of fading evolution
5.2.1. Distribution of random power series in cases of uniform and Erlang distributions
5.2.2. The distribution of the limiting position
5.3. Differential and integral equations for jump random motions
5.3.1. The Erlang jump telegraph process on a line
5.3.2. Examples

5.4. Estimation of the number of level crossings by the telegraph process
5.4.1. Estimation of the number of level crossings for the telegraph process in Kac's condition

References

Index

Other titles from

iSTE

in

Mathematics and Statistics

2020

BARBU Vlad Stefan, VERGNE Nicolas
Statistical Topics and Stochastic Models for Dependent Data with Applications

CHABANYUK Yaroslav, NIKITIN Anatolii, KHIMKA Uliana
Asymptotic Analyses for Complex Evolutionary Systems with Markov and Semi-Markov Switching Using Approximation Schemes

KOROLIOUK Dmitri
Dynamics of Statistical Experiments

MANOU-ABI Solym Mawaki, DABO-NIANG Sophie, SALONE Jean-Jacques
Mathematical Modeling of Random and Deterministic Phenomena

2019

BANNA Oksana, MISHURA Yuliya, RALCHENKO Kostiantyn, SHKLYAR Sergiy
Fractional Brownian Motion: Approximations and Projections

GANA Kamel, BROC Guillaume
Structural Equation Modeling with lavaan

KUKUSH Alexander
Gaussian Measures in Hilbert Space: Construction and Properties

LUZ Maksym, MOKLYACHUK Mikhail
Estimation of Stochastic Processes with Stationary Increments and Cointegrated Sequences

MICHELITSCH Thomas, PÉREZ RIASCOS Alejandro, COLLET Bernard, NOWAKOWSKI Andrzej, NICOLLEAU Franck
Fractional Dynamics on Networks and Lattices

VOTSI Irene, LIMNIOS Nikolaos, PAPADIMITRIOU Eleftheria, TSAKLIDIS George
Earthquake Statistical Analysis through Multi-state Modeling
(Statistical Methods for Earthquakes Set – Volume 2)

2018

AZAÏS Romain, BOUGUET Florian
Statistical Inference for Piecewise-deterministic Markov Processes

IBRAHIMI Mohammed
Mergers & Acquisitions: Theory, Strategy, Finance

PARROCHIA Daniel
Mathematics and Philosophy

2017

CARONI Chysseis
First Hitting Time Regression Models: Lifetime Data Analysis Based on Underlying Stochastic Processes
(Mathematical Models and Methods in Reliability Set – Volume 4)

CELANT Giorgio, BRONIATOWSKI Michel
Interpolation and Extrapolation Optimal Designs 2: Finite Dimensional General Models

CONSOLE Rodolfo, MURRU Maura, FALCONE Giuseppe
Earthquake Occurrence: Short- and Long-term Models and their Validation
(Statistical Methods for Earthquakes Set – Volume 1)

D'AMICO Guglielmo, DI BIASE Giuseppe, JANSSEN Jacques, MANCA Raimondo
Semi-Markov Migration Models for Credit Risk
(Stochastic Models for Insurance Set – Volume 1)

GONZÁLEZ VELASCO Miguel, del PUERTO GARCÍA Inés, YANEV George P.
Controlled Branching Processes
(Branching Processes, Branching Random Walks and Branching Particle Fields Set – Volume 2)

HARLAMOV Boris
Stochastic Analysis of Risk and Management
(Stochastic Models in Survival Analysis and Reliability Set – Volume 2)

KERSTING Götz, VATUTIN Vladimir
Discrete Time Branching Processes in Random Environment
(Branching Processes, Branching Random Walks and Branching Particle Fields Set – Volume 1)

MISHURA YULIYA, SHEVCHENKO Georgiy
Theory and Statistical Applications of Stochastic Processes

NIKULIN Mikhail, CHIMITOVA Ekaterina
Chi-squared Goodness-of-fit Tests for Censored Data
(Stochastic Models in Survival Analysis and Reliability Set – Volume 3)

SIMON Jacques
Banach, Fréchet, Hilbert and Neumann Spaces
(Analysis for PDEs Set – Volume 1)

2016

CELANT Giorgio, BRONIATOWSKI Michel
Interpolation and Extrapolation Optimal Designs 1: Polynomial Regression and Approximation Theory

CHIASSERINI Carla Fabiana, GRIBAUDO Marco, MANINI Daniele
Analytical Modeling of Wireless Communication Systems
(Stochastic Models in Computer Science and Telecommunication Networks Set – Volume 1)

GOUDON Thierry
Mathematics for Modeling and Scientific Computing

KAHLE Waltraud, MERCIER Sophie, PAROISSIN Christian
Degradation Processes in Reliability
(Mathematial Models and Methods in Reliability Set – Volume 3)

KERN Michel
Numerical Methods for Inverse Problems

RYKOV Vladimir
Reliability of Engineering Systems and Technological Risks
(Stochastic Models in Survival Analysis and Reliability Set – Volume 1)

2015

DE SAPORTA Benoîte, DUFOUR François, ZHANG Huilong
Numerical Methods for Simulation and Optimization of Piecewise Deterministic Markov Processes

DEVOLDER Pierre, JANSSEN Jacques, MANCA Raimondo
Basic Stochastic Processes

LE GAT Yves
Recurrent Event Modeling Based on the Yule Process
(Mathematical Models and Methods in Reliability Set – Volume 2)

2014

COOKE Roger M., NIEBOER Daan, MISIEWICZ Jolanta
Fat-tailed Distributions: Data, Diagnostics and Dependence
(Mathematical Models and Methods in Reliability Set – Volume 1)

MACKEVIČIUS Vigirdas
Integral and Measure: From Rather Simple to Rather Complex

PASCHOS Vangelis Th
Combinatorial Optimization – 3-volume series – 2nd edition
Concepts of Combinatorial Optimization / Concepts and Fundamentals – volume 1
Paradigms of Combinatorial Optimization – volume 2
Applications of Combinatorial Optimization – volume 3

2013

COUALLIER Vincent, GERVILLE-RÉACHE Léo, HUBER Catherine, LIMNIOS Nikolaos, MESBAH Mounir
Statistical Models and Methods for Reliability and Survival Analysis

JANSSEN Jacques, MANCA Oronzio, MANCA Raimondo
Applied Diffusion Processes from Engineering to Finance

SERICOLA Bruno
Markov Chains: Theory, Algorithms and Applications

2012

BOSQ Denis
Mathematical Statistics and Stochastic Processes

CHRISTENSEN Karl Bang, KREINER Svend, MESBAH Mounir
Rasch Models in Health

DEVOLDER Pierre, JANSSEN Jacques, MANCA Raimondo
Stochastic Methods for Pension Funds

2011

MACKEVIČIUS Vigirdas
Introduction to Stochastic Analysis: Integrals and Differential Equations

MAHJOUB Ridha

Recent Progress in Combinatorial Optimization – ISCO2010

RAYNAUD Hervé, ARROW Kenneth
Managerial Logic

2010

BAGDONAVIČIUS Vilijandas, KRUOPIS Julius, NIKULIN Mikhail
Nonparametric Tests for Censored Data

BAGDONAVIČIUS Vilijandas, KRUOPIS Julius, NIKULIN Mikhail
Nonparametric Tests for Complete Data

IOSIFESCU Marius *et al.*
Introduction to Stochastic Models

VASSILIOU PCG
Discrete-time Asset Pricing Models in Applied Stochastic Finance

2008

ANISIMOV Vladimir
Switching Processes in Queuing Models

FICHE Georges, HÉBUTERNE Gérard
Mathematics for Engineers

HUBER Catherine, LIMNIOS Nikolaos *et al.*
Mathematical Methods in Survival Analysis, Reliability and Quality of Life

JANSSEN Jacques, MANCA Raimondo, VOLPE Ernesto
Mathematical Finance

2007

HARLAMOV Boris
Continuous Semi-Markov Processes

2006

CLERC Maurice
Particle Swarm Optimization

Printed and bound by CPI Group (UK) Ltd, Croydon, CR0 4YY

15/09/2026

14962362-0002